Particulate Matter: Properties and effects upon health

Particulate Matter: Properties and effects upon health

Edited by

R.L. Maynard[a] and C.V. Howard[b]

[a]Department of Health, Skipton House, 80 London Road, London SE1 6LW, UK

[b]Fetal and Infant Toxico-Pathology, University of Liverpool, Liverpool L69 7ZA, UK

CRC Press
Taylor & Francis Group
Boca Raton London New York

CRC Press is an imprint of the
Taylor & Francis Group, an **informa** business
A TAYLOR & FRANCIS BOOK

Front cover: Sonicated CB M120 exhibiting a compact arrangement of spherulites (see *Figure 4.6*).

Published 1999 by CRC Press
Taylor & Francis Group
6000 Broken Sound Parkway NW, Suite 300
Boca Raton, FL 3487-2742

ISBN-13: 978-1-85996-172-8 (hbk)

Typeset by Marksbury Multimedia Ltd, Midsomer Norton, Bath, UK.

Visit the Taylor & Francis Web site at
http://www.taylorandfrancis.com

and the CRC Press Web site at
http://www.crcpress.com

Publisher's Note
The publisher has gone to great lengths to ensure the quality of this reprint but points out that some imperfections in the original may be apparent

A CIP catalogue record for this book is available from the British Library.

Contents

Abbreviations

AI	alveolar interstitial
APEG	Airborne Particles Expert Group
BAL	bronchoalveolar lavage
BB	bronchial
bb	bronchiolar
CB	carbon black
COMEAP	UK Committee on the Medical Effects of Air Pollutants
CVS	constant volume sampling
cwp	coal workers' pneumoconiosis
DEP	diesel exhaust particles
DETR	Department of Environment, Transport and Regions
DI	direct injection
ELPI	electrical low-pressure impactor
EPEFE	European Programme on Emissions, Fuels and Engine Technologies
EPXMA	electron probe X-ray microanalysis
ESD	equivalent spherical diameter
ET	extrathoracic
FESEM	field emission scanning electron microscopy
GC-MS	gas chromatography mass spectrometry
GGT	gamma glutamyl transpeptidase
IA	image analysis
IC	ion chromatography
ICP-AES	inductively coupled plasma atomic emission spectroscopy
ICP-MS	inductively coupled plasma mass spectrometry
IDI	indirect injection
IP	Institute of Petroleum
MMMFs	man-made mineral fibres
MTD	maximum tolerated dose
NAQS	National Air Quality Strategy
PAH	polyaromatic hydrocarbons
PORG	Photochemical Oxidants Review Group
PTFE	polytetrafluoroethylene
QCM	quartz crystal microbalance
QUARG	Quality of Urban Air Review Group
RGAR	Review Group on Acid Rain
ROFA	residual oil fly ash

SMPS	scanning mobility particle sizer
SOD	superoxide dismutase
SOF	soluble organic fraction
TEM	transmission electron microscopy
TEOM	tapered element oscillating microbalance
TNF	tumour necrosis factor
UF	ultrafine
VOF	volatile organic fraction
WHO	World Health Organization
XRD	X-ray diffraction

Contributors

BéruBé, K., School of Biosciences, Cardiff University, Biomedical Buildings, PO Box 911, Museum Avenue, Cardiff, CF1 3US, UK

Davies, M.J., Rover Group, Building 507a, Gaydon Test Centre, Banbury Road, Lighthorne, Warwick, CV34 0RG, UK

Donaldson, K., Biomedical Research Group, Napier University, 10 Colinton Road, Edinburgh, EH10 5DT, UK and Edinburgh Lung and the Environment Group Initiative, Colt Research Laboratories Medical School, Teviot Place, Edinburgh, UK

Harrison, R.M., Institute of Public and Environmental Health, University of Birmingham, Edgbaston, Birmingham, B15 2TT, UK

Jefferson, D., Department of Chemistry, Cambridge University, Lensfield Road, Cambridge, CB2 1EW, UK

Jones, T.P., School of Biosciences, Cardiff University, Biomedical Buildings, PO Box 911, Museum Avenue, Cardiff, CF3 1US, UK

MacNee, W., Edinburgh Lung and the Environment Group Initiative, Colt Research Laboratories Medical School, Teviot Place, Edinburgh, UK

Masek, L., CRC Wessex Regional Medical Oncology Unit, University of Southampton, Southampton General Hospital, Southampton SO16 6YD, UK

McAughey, J., AEA Technology Environment, E6/1.04 Culham, Abingdon, OX14 3ED, UK

Murphy, S.A., School of Biosciences, Cardiff University, PO Box 911, Museum Avenue, Cardiff, CF1 3US, UK

Pooley, F., Cardiff School of Engineering, Cardiff University, Queens Buildings, PO Box 685, The Parade, Cardiff, CF2 3TA, UK

Richards, R.J., School of Biosciences, Cardiff University, Biomedical Buildings, PO Box 911, Museum Avenue, Cardiff, CF1 3US, UK

Seaton, A., Environmental and Occupational Medicine, University of Aberdeen, University Medical School, Foresterhill, Aberdeen, AB9 22D, UK

Stone, V., Biomedicine Research Group, Napier University, 10 Colinton Road, Edinburgh EH10 5DT, UK

Symons, D., School of Biosciences, Cardiff University, Biomedical Buildings, PO Box 911, Museum Avenue, Cardiff, CF1 3US, UK

Williams, M.L., Department of Environment, Transport and the Regions, Air and Environment Quality Division, Ashdown House, Zone 4E14, 123 Victoria Street, London, SW1E 6DE, UK

Williamson, B.J., Department of Earth Sciences, University of Bristol, Wills Memorial Building, Queen's Road, Bristol BS8 1RJ, UK

Foreword

Foreword

C. Vyvyan Howard

Background

In 1986–1987, as one of the tasks given to me during my Presidency of the Royal Micoscopical Society (RMS), I was charged with finding a topic which would unite all the different disciplines and strands of activity in RMS within a single session at the scientific meeting Micro '98. Particulate aerosols provided the answer and grew from an idea seeded by listening to a paper presented by Dr David Jefferson at Micro '96, on the physical chemistry and crystallography of nanometric particles. The study of miniscule particles, often consisting of only a few atoms, will inevitably at some stage involve the use of microscopes. Nanoparticles interest physicists and physical chemists who are pushing high resolution electron microscopes to the limits. They interest surface scientists who study the catalytic properties of nano-particles. They interest biologists and pathologists who want to know the possible effects of this chemistry upon *in vivo* aqueous systems. In addition to these disciplines we were able to attract scientists with interests in atmospheric chemistry, meteorology, vehicle engine design and a number of other activities. This, the resulting book, amounts to a very representative snapshot in time of our current state of knowledge of the field.

In fact because of its multidisciplinary nature the RMS proves to be an ideal forum for a debate on particulate matter. At the atomic level high resolution scanning transmission electron microscopy (STEM), X-ray diffraction techniques and scanning probe microscopies (SPM) give insights into the compositional nature and surface characteristics of nano-particles. At lower resolution the latest advances in light microscopy, including the study of living tissues, can be brought to bear on the study of pathological processes. It transpires that my own field of quantitative toxico-pathology (Howard and Reed, 1998) has already been applied to the estimation of the number of experimentally inhaled particles in lung and their clearance rate (Geiser *et al.*, 1990, 1994, 1995; Maye *et al.*, 1997) and can be expected to contribute more in the future.

The basic physical chemistry of nano-particles

For the general reader one of the more technically demanding presentations in this book is Chapter 5 by Jefferson and Tilley on "The Structural and Physical Chemistry of Nanoparticles". However it is, in my opinion, a crucial contribution and points the way forward for much of the research that urgently needs to be performed. It transpires that materials which are regarded as being chemically relatively inert can become highly reactive and electrically charged when transformed into small particles. This comes as no surprise to those involved in the manufacture of catalysts, where nano-particle production is often part of the process. Not only does atomisation of materials lead to an increase in chemical reactivity but also, on occasion, to the production of completely novel configurations that cannot exist in the bulk material from which the particles originated. The study of the chemistry of such novel configurations is in its infancy.

The effect of mixtures of particles of differing chemical composition entering the blood stream in large numbers on a daily basis, is only just beginning to be investigated. Clearly the surface chemistry of nano-particles and their effect on organic molecules in an aqueous environment is going to be difficult but very important to study and will need an interdisciplinary approach.

Are particulate aerosols harmful to health?

There is no doubt that some particulate aerosols are indeed hazardous. However, the degree of hazard associated with specific types of particle and the precise mechanisms by which exposure leads to pathology are as yet poorly understood and currently the subject of increasingly intense research.

However, there is a group in society who are regularly and voluntarily exposed to inhaled particulate aerosols, namely cigarette smokers. By examining what is known of the patterns of pathology in smokers and then comparing with the epidemiological and experimental findings reported by contributors to this book, there are patterns to be discerned.

We know that some smokers, but by no means all, will develop cancers in association with their habit. What has also been known, from a number of well designed scientific investigations, is that smokers will virtually all damage their arterial systems, in a manner dependent on the 'dose' of cigarettes that they consume. Auerbach *et al.* (1965) studied 1372 men who died from causes other than coronary heart disease between 1 January 1959 and 31 December 1960 at the Veterans Administration Hospital, East Orange, New Jersey. They assessed the coronary arteries into four categories for degree of atheroma. Several weeks after autopsy experienced interviewers, blinded to the outcome of the autopsy, established the smoking habits of the study subjects which were also classified into four categories. *Figure 1* reports some of the results of the study. It clearly shows the positive correlation between the quantity of cigarettes consumed and the severity of atherosclerosis in coronary arteries.

The message seems very clear, if you smoke cigarettes there is an overwhelming likelihood that you will damage your coronary arteries.

How does this fit into what is known about human exposure to environmental particulate aerosols? A number of contributors to this book have drawn attention to large scale epidemiological studies that

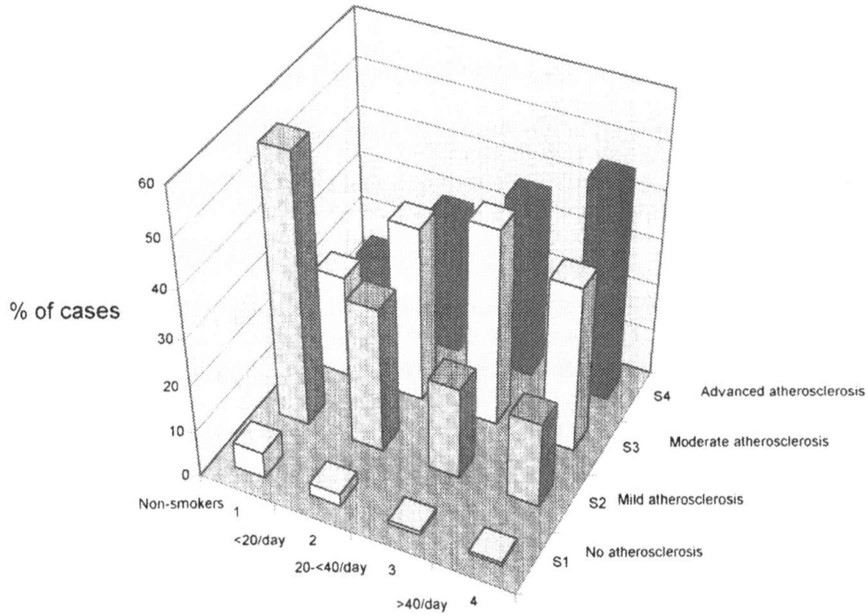

Figure 1. Degree of atherosclerosis in coronary arteries related to smoking habits. This 3-D plot is constructed from data in *Table 3* of Auerbach *et al.* (1965). The *x*-axis is divided into four categories of cigarette consumption. The *y*-axis is divided into four categories of postmortem outcome with respect to coronary artery atherosclerosis. The *z*-axis reports the percentage of cases within each smoking class with respect to coronary artery condition.

have been conducted, for example Dockery *et al.* (1993), which suggest that there may be a relationship between PM_{10} exposure and cardiovascular disease. These results are disputed by some. However, Seaton *et al.* (1995) have suggested a mechanism as to how PM_{10} exposure might cause morbidity. Donaldson, Stone and MacNee in Chapter 8 bring forward evidence for an increase in the coagulability of the blood upon exposure to PM_{10} aerosols, suggesting that free radicals might be the culprits. Transitional metals in PM_{10} aerosols have been associated with their pro-inflammatory effects, which draws interesting associations with some of the findings presented by Jefferson and Tilley in Chapter 3.

Tobacco smoke is a very complex mixture but it certainly contains a PM_{10} fraction which will also have free-radicals present. There may be other parallels that are not yet obvious. However, we can at least appreciate that there is common ground between the known pathologies associated with smoking and the hypothesized damage caused by chronic particulate aerosol inhalation. This therefore lends credance to the mechanistic hypotheses and toxicological findings to date in association with environmental particulate pollution.

Where does this book lead us?

Like most good science, the questions posed by the findings presented here far outnumber the solutions. It is clear that we need to understand much more about the interaction between idiosynchrasies of nanoparticle chemistry and normal biochemistry. That was generally agreed in the open forum discussion following the formal presentations at Micro '98. Only when this is further elucidated is there a realistic chance of approaching questions such as "What types of particle are the most dangerous?", answers to which are necessary before we can approach other questions such as "Where do we put the maximum effort in reducing PM_{10} aerosols?".

In a society which depends heavily upon combustion processes for power production, transport and heavy industry, a world conceived without anthropogenic particulate aerosols is unrealistic and unobtainable in the forseeable future. Further objective multidisciplinary research will aid decision makers in their task of optimizing cost and benefit. It may prove to be that some processes contribute a disproportionate amount of the toxic burden attributable to the PM_{10} fraction. For example municipal waste incinerators, which have appreciable heavy and transitional metal contents in their feedstocks, are likely to be more damaging than gas fired power stations, for each unit of energy produced. Yet there does not appear to have been any systematic analysis of respective PM_{10} efluvia of different industries with respect to the basic physical chemistry or the types of health effects that have been discussed in this book. This should become an urgent aim for research.

On the other hand some contributions to PM_{10} aerosols are part of Nature and cannot be controlled. For example, as Roy Harrison discusses in Chapter 9, up to 30% of the PM_{10} particles that we breath in UK air can come from the action of the sea producing suspended salt crystals (marine aerosol). This is clearly a process that has been in action since time immemorial and, with naturally occurring salts, presumably for the most part non-toxic. However, it raises questions over the past and continuing practice of using coastal waters for the disposal of, for example radionuclides with long half-lives, which are known to become resuspended in air (Eakins and Lally, 1984).

Another example of a dilemma faced by regulators is that of vehicle emissions, as discussed by many contributors to this book. Currently much research is being directed at ways of decreasing emmissions from internal combustion engines. The use of particle traps, discussed by Michael Davis in Chapter 10, with secondary combustion is being developed. This will, in turn, drive the size distribution of particles in exhausts towards smaller sizes, which might or might not be beneficial. In fact a recent paper (Rudell *et al.*, 1999) has shown that exposure to particle-trap filtered diesel exhaust did not significantly reduce bronchoalveolar inflammation when compared to exposure to unfiltered diesel exhaust. Again, analyses of the surface chemistry of such particles are to be encouraged, in association with toxicological studies. The trend towards the use of shorter chain-length fuels, such as LPG, is already receiving support from government and may prove to be an easier route. However, the need to develop reliable and reproducible bioassays, both *in vivo* and *in vitro*, in association with relevant chemical analyses is urgent if the regulators are to be assisted in making rational decisions concerning these problems.

In conclusion

It has been both enjoyable and stimulating to be associated with the organization of the meeting at Micro '98 and the subsequent production of this book. I take this oportunity to offer my personal thanks and that of the Royal Microscopical Society to all the contributors for their time and effort and in particular my co-editor Dr Bob Maynard from the Department of Health. The cumulative expertise contained between these covers is enormous. I hope that the book will have the effect of increasing the level of collaboration between the different disciplines involved and also spawn some new ones. There are a number of techniques in microscopy not covered in this volume but which would be expected to have valuable contributions to make. These include 'Super-STEM' microscopy, scanning tunelling microscopy, atomic force microscopy and scanning proton microscopy, to mention but four. There is no doubt that microscopy will remain a central plank in the research effort to increase our undertanding of how particulate aerosols affect our health.

C. Vyvyan Howard
Liverpool, August 1999

References

Auerbach, O., Cuyler Hammond, E. and Garfinkel, L. (1965). Smoking in relation to atherosclerosis of the coronary arteries. *New England J. Med.* **273:** 775–779.

Dockery, D.W., Pope, C.A., Xu, X.P., Spengler, J.D., Ware, J.H., Fay, M.E., Ferris, B.G. and Speizer, F.E. (1993). An association between air-pollution and mortality in 6 united-states cities. *New England J. Med.* **329:** 1753–1759.

Eakins, J.D. and Lally, A.E. (1984). The transfer to land of actinide bearing sediments from the Irish Sea by spray. *Sci. Total Environ.* **35:** 23–32.

Geiser, M., Serra, A.L., Cruz Orive, L.M., Baumann, M., Hof, V.I. and Gehr, P. (1995). Efficiency of airway macrophage recovery by bronchoalveolar lavage in hamsters – a stereological approach. *Europ. Resp. J.* **8:** 1712–1718.

Geiser, M., Baumann, M., Cruz Orive, L.M., Hof, V.I., Waber, U. and Gehr, P. (1994). The effect of particle inhalation on macrophage number and phagocytic activity in the intrapulmonary conducting airways of hamsters. *Am. J. Resp. Cell Mol. Biol.* **10:** 594–603.

Geiser, M., Cruz, Orive, L.M., Hof, V.I. and Gehr, P. (1990) Assessment of particle retention and clearance in the intrapulmonary conducting airways of hamster lungs with the fractionator. *J. Microscopy* **160:** 75–88.

Howard, C.V. and Reed, M.G. (1998) *Unbiased Stereology, Three-dimensional Measurement in Microscopy.* BIOS Scientific Publishers, Oxford.

Maye, I., deFraissinette, A., Cruz Orive, L.M., Vonderscher, J., Richter, F. and Gehr, P. (1997) Comparison of the rate of phagocytosis of orthorhombic cyclosporine A (CsA) and latex particles by alveolar macrophages from hamsters. *Cellular Mol. Life Sci.* **53:** 689–696.

Rudell, B., Blomberg, A., Helleday, R., Ledin, M.-C., Lundback, B., Stjernberg, N., Horstedt, P. and Sandstrom, T. (1999). Bronchoalveolar inflammation after exposure to diesel exhaust: comparison between unfiltered and particle trap filtered exhaust. *Occup. Environ. Med.* **56**: 527–534.

Seaton, A., MacNee, W., Donaldson, K. and Godden, D. (1995). Particulate air pollution and acute health effects. *Lancet* **345:** 176–178.

1 Introduction

Robert L Maynard

The views given here are those of the author and should not be taken as those of the UK Department of Health.

Air pollution is not a new phenomenon. Pollution of the air by particles is also not a new phenomenon; indeed, of the commonly studied air pollutants smoke, consisting of course of a particulate aerosol, was the first pollutant to be complained of and studied in detail. It is usual to trace the history of complaints about smoke in the UK to the 13th century when the use of coal for domestic heating and the firing of furnaces produced severe air pollution in London (Brimblecombe, 1987). Though there have probably been significant fluctuations in the degree of pollution, the problem of particles has persisted in London for 700 years. The lack of measurements of concentrations of particles in the air make it difficult to say when pollution in London was at its worst. It was certainly bad in the 17th century when the diarist John Evelyn published his *Fumifugium or the Inconvenience of the Aer and Smoake of London Dissipated* in 1661, and again in the 19th century when Charles Dickens described with characteristic detail, the smoke and fogs of London and other British cities and towns. Cold weather was noticed to be associated with fogs and the exceptionally cold winters of the last decade of the 19th century were associated with dense fogs in London. The combination of temperature inversions and a source, domestic chimneys, of unusual strength, led to the fogs being heavily laden with smoke.

In 1905 Des Voeux, Secretary to the Coal Smoke Abatement Society (later the National Society for Clean Air), coined the word smog to describe this combination of smoke and fog. The word has been used more recently to describe the products of photochemical reactions: ozone, aldehydes and secondary particles, as encountered in cities such as Los Angeles. It should be noted that the London smog that Des Voeux was describing was a reducing rather than an oxidizing chemical mixture and

Particulate Matter, edited by R.L. Maynard and C.V. Howard

contained high concentrations of smoke particles and droplets of sulphur dioxide. This 'old fashioned' smog still exists in those parts of the world where coal is still burned on a large scale.

Despite efforts to reduce the production of smoke in London conditions remained poor and in the early decades of the 20th century, and in 1948 and 1952, severe smogs occurred. The 1952 London episode, 'The Great Smog', was for many people literally the last straw, and some 4000 extra deaths occurred as a result (Ministry of Health, 1954). A disaster of such proportions could not be ignored and the Government, adopting a Private Member's Bill advanced by the MP Gerald Nabarro, passed the Clean Air Act in 1956. This Act, the first of its kind, provided reimbursement to Local Authorities which established smoke-free zones and made grants to citizens to enable them to convert their domestic open fires to grates suitable for the combustion of smokeless fuel. The Act was successful and, as the smokeless zones spread, cities emerged from the winter gloom of centuries. True, the blackened buildings still bore witness to the past conditions, but gradually, clean-up work was undertaken and buildings such as the Houses of Parliament and Westminster Cathedral returned to their original colours. It is difficult, today, to find an example of a truly black public building: the memorial to Sir Walter Scott in Princes' Street, Edinburgh, remains black and thus a memorial, also, to the days when Edinburgh was dubbed 'Auld Reekie'.

The London smog of 1952 also triggered research into the effects of air pollutants on health. This had previously been patchy and work had focused on episodes of pollution such as those which occurred in the Meuse Valley in Belgium in 1930, and in Donora, Pennsylvania, USA in 1948 (Firket, 1936; Schrenk *et al.*, 1949). Work at the Medical Research Council's Air Pollution Research Unit at St Bartholomew's Hospital, under the Directorship of Professor P. J. Lawther, led the way. He, with colleagues, studied not only the effects on health of episodes of pollution, but also the pathophysiology of the effects of the main pollutants and the effects of day-to-day variations in levels of pollution on the health of patients (Waller, 1991). These patients were drawn from a clinic for those suffering from chronic bronchitis. Just as smog characterized the great cities of the UK, so did chronic bronchitis, 'The British Disease' characterize the old age of so many of her citizens. That the prevalence of this disease might be in some way linked with exposure to pollution did not escape Lawther and he argued that the effects of long-term exposure were as deserving of study as the effects of severe air pollution episodes (Lawther, 1961). This call has still not been adequately heeded. It was also in the Air Pollution Unit that Mr Robert Waller first applied the techniques of electron microscopy to the examination of the particles found in London air (Waller *et al.*, 1963). Estimates of the number of particles found in unit volumes of air were produced: these proved to be remarkably accurate. Modern electron-microscopic techniques are now being applied to the problem. These are discussed by Professor Pooley and Dr BéruBé in Chapters 3 and 4.

As levels of air pollution fell in London so it became more and more difficult to detect the effects on health of such day-to-day variations in concentrations of smoke and sulphur dioxide as still occurred. This led to the identification of thresholds of effect: the view being that as long as the smoke concentration remained below 250 $\mu g\ m^{-3}$ and the sulphur dioxide concentration below 500 $\mu g\ m^{-3}$ (24-h average concentrations) such patients as had been studied would be unlikely to detect effects upon their health as a result of day-to-day fluctuations in concentrations. This work was seminal in establishing guidelines for air quality and played a large part in providing a basis for the World Health Organization's Air Quality Guidelines for Europe 1987. The figures provided by these studies stood as accepted thresholds until the early 1990s (WHO, 1990). Then things changed.

A trickle of new epidemiological studies that began in the late 1980s and turned into a flood in the 1990s, provided evidence that day-to-day variations in the already low concentrations of particles and other pollutants were still associated with effects on health. A number of reviews of this literature have been published (Department of Health, 1995; Lambert *et al.*, 1998; Wilson and Spengler, 1996). These effects included increases in daily numbers of deaths and hospital admissions and less severe effects such as visits to general practitioners, symptoms and the consumption of anti-asthma remedies. To say that these studies have provoked controversy would be to seriously understate the case. Vigorous and sometimes acrimonious debate followed both in the correspondence columns of learned journals and in a less restrained form on the floors of international symposia. Critics argued that the results were artefactual and contrary to both common sense and established doctrine. Why were the arguments so ill-tempered? Three reasons can be suggested:

- the results of the new studies were so significantly different from those of the older studies that it seemed to some that if the new results were right then the old results had to be wrong. The results of the older studies had been reflected upon for years and were generally considered to be sound, though doubts had been expressed;
- toxicologists wondered how such small doses of pollutants as implied by the results of the new studies, could produce any effects, let alone death, and felt that the epidemiological studies must be flawed;
- those who represented, or who were associated with, industries and manufacturers that produced pollutants or devices such as motor vehicles that emit pollutants, saw that demands for cleaner operations and products would follow and realized that these would be both difficult to meet and inevitably expensive.

It is fair to say that the results of the new epidemiological studies are still not accepted by all as indicative of causal associations between low concentrations of pollutants and effects on health. Of the several pollutants studied, most emphasis has been placed on the results regarding particles. Several well argued and detailed papers have been published disputing the interpretation of these studies and have argued that the reported associations are produced, perhaps, by inadequate accounting for confounding factors and should not be taken as indicating a causal relationship (Gamble, 1998; Gamble and Lewis, 1996; Lipfert and Wyzga, 1995). The case cannot be argued again here. It might just be said that a number of detailed examinations of the evidence have been undertaken by independent experts: their conclusions have been similar and have suggested that reported associations are important and should be regarded as causal. This view has been accepted by the UK Department of Health. This acceptance of the evidence has moved the debate forward to a search for an explanation of the effects. This has not proved easy to find and, though several hypotheses have been produced, none has been accepted (Oberdörster *et al.*, 1995; Seaton *et al.*, 1995).

The sticking point for many toxicologists can be summed up simply: effects are said to occur when the daily dose of particles increases by as little as 250 μg; airborne particles do not contain any extraordinarily toxic substances and, therefore, such a dose could not have much effect, certainly could not kill anybody, and therefore the original evidence is flawed. This has been put deliberately starkly.

One of the most interesting ideas put forward to circumvent this objection has been that it is not the chemistry of the particulate aerosol but, rather, its physical properties that are important. This has shifted the debate into a new direction. It was suggested that very small particles, of nanometre dimensions, might be playing a role. This has been supported by studies that showed that materials such a titanium dioxide and carbon black had very different, and more marked, toxicological properties when presented as ultrafine (<100 nm diameter) particles than as particles of about 250 nm diameter (Amdur *et al.*, 1988; Gilmour *et al.*, 1997; Li *et al.*, 1996; Oberdörster *et al.*, 1992). Such a theory is appealing as it can make sense of the similar results produced by epidemiological studies undertaken in areas between which there are significant differences in the chemical composition of particles. For example, studies in European cities will largely reflect the effects of a primary aerosol produced by motor vehicles; studies in rural areas, on the other hand, might largely reflect the effects of a secondary aerosol produced by photochemical reactions. The aerosols differ significantly in chemical composition but each contains many ultrafine particles. How such ultrafine particles act remains unknown.

A point that has been ignored until recently is the capacity of ultrafine particles to be retained in the lungs. It was often assumed that such particles would simply waft in and out of the respiratory system in the breath. Nothing could be further from the truth. Recent modelling studies have shown that some 60% of inhaled particles of 20 nm diameter are deposited in the gas exchange zone of the lung (ICRP, 1994). Interestingly, it is true that few even smaller particles (<10 nm diameter) are retained in the lung: this is because they do not reach the lung but are essentially all deposited in the nose and pharynx.

A recurring question regarding the possible effects of ambient concentrations of particles has been, if man (and other animals) have always been exposed to particles, why have we (and they) not evolved a defence against them? We should ask in return whether particles of nanometre dimensions have always been common. Aerosols generated by physical means certainly do not contain many such small particles – it takes a great deal of energy to reduce, say, a rock to ultrafine particles. Fires, on the other hand, generate prodigious numbers of ultrafine particles: a coal gas flame burning for 15 s can raise the number count of particles in a chamber from 109 000 to 860 000 per cm^3 (Gadle and Magill, 1956). These small particles, 'nuclei', tend to be short lived and rapidly aggregate to produce particles of 0.5–2.5 μm diameter. It is interesting to note that particles of <0.5 μm diameter are minimally deposited in the respiratory system: the physics of diffusion and sedimentation conspiring against their deposition.

Particles deposited in the gas exchange zone of the lung are removed by macrophages. Our knowledge of how macrophages recognize particles has increased dramatically during the past few years (Kobzik, 1995). Despite this, we lack information on how macrophages recognize, ingest and handle ultrafine particles. We do know that macrophages can be overloaded with particles and that they then lose their capacity to migrate and leave the gas exchange zone. Recent studies have suggested that overloading with very small particles may occur at lower percentage volume loadings than with larger particles (Donaldson *et al.*, 1998). This point is taken up by Professor Donaldson in Chapter 8. Donaldson also takes up the interesting property of ultrafine particles of generating free radicals. One of the more remarkable recent discoveries in the area of particle toxicology has been that of the capacity of ultrafine particles of substances such as carbon black to generate hydroxyl radicals (Li *et al.*, 1996). The exact mechanism by which such radicals are generated is unknown but it is possible that, at least in some cases, transition metals such as ferrous iron may play a catalytic role.

A great deal is known by chemists about the catalytic properties of nanometre dimension particles of iron and other metals. This information has not, generally, penetrated into toxicological thinking and one of the purposes of the Symposium was to draw together workers from both disciplines. Dr Jefferson has provided a detailed discussion of the chemical aspects of this form of catalysis in Chapter 5.

Research into the effects of particles on health has put pressure on those with the responsibility for regulating air quality. Developing regulatory strategies tends to be a slow process though recent years have seen a dramatic acceleration in a number of countries. The rate of development of thinking in the area of particle toxicology and epidemiology has posed problems and the regulator might remember the Red Queen's observation to Alice: 'It takes all the running *you* can do to keep in the same place'. This has meant that regulatory thinking has sometimes seemed behind the leading edge of the science. This has been especially the case regarding the optimal metric that should be chosen to reflect the concentration of particles in the air. In the UK, and a number of other countries, attention has focused on PM_{10} (mass of particles generally less than 10 μm aerodynamic diameter) but pressure to move to $PM_{2.5}$ or even $PM_{1.0}$ is mounting. It may be, though this is by no means certain, that a metric based on particle mass is less than ideal and that regulations should focus on controlling the number of ultrafine particles in the air. Dr Williams has provided a thoughtful discussion of the problems facing the regulator in Chapter 11.

Air pollution will not go away and resources for dealing with air pollution are not finite: these are certainties. Less certain, but probable, is that public concern about air pollution and its effects will not abate. These factors mean that efforts to control air pollution will need to be maintained and focused upon those measures that are most likely to provide a good return in terms of benefit as balanced against cost. Controlling levels of particles is likely to be one of the measures that will be pressed for strongly. Unless the toxicologically active components, chemical or physical of the ambient aerosol are known and understood in some detail, efforts to achieve control may be misdirected. For example, it is conceivable that techniques designed to reduce the emissions of particle mass by motor vehicles could increase the emissions in terms of numbers of ultrafine particles. If these particles are the key to effects on health such efforts would be counter-productive. The emphasis of this symposium on ultrafine particles is thus well directed and a wide range of scientific approaches will be needed to solve the difficult problems in this area. The presentations given at the Symposium show the range of relevant activities in the area in the UK: it is encouraging that the UK can field such expertise in such an important area.

References

Amdur, M.O., Chen, L.C., Guty, J., Lam, H.F. and Miller, P.D. (1988) Speciation and pulmonary effects of acidic SO_x formed on the surface of ultrafine zinc oxide aerosols. *Atmospheric Environ.* **22:** 557–560.

Brimblecombe, P. (1987) *The Big Smoke: a History of Air Pollution in London Since Medieval Times*. Methuen, London.

Cadle, R.D. and Magill, P.L. (1956) Chemistry of contaminated atmospheres. In: *Air Pollution Handbook. Section 3* (eds P.L. Magill, F.R. Holden and C. Ackley). McGraw Hill Book Company, New York, pp. 3–1 to 3–27.

Department of Health. Committee on the Medical Effects of Air Pollutants (1995) *Non-biological Particles and Health*. HMSO, London.

Donaldson, K., Li, X.Y. and MacNee, W. (1998) Ultrafine (nanometre) particle associated lung injury. *J. Aerosol Sci.* **29:** 553–560.

Evelyn, J. (1961) *Fumifugium or the Inconvenience of the Aer and Smoake of London Dissipated* (Reprint). First published 1661. National Society for Clean Air, Brighton, UK.

Firket, J. (1936) Fog along the Meuse Valley. *Trans. Faraday Soc.* **32:** 1192–1197.

Gamble, J.F. (1998) $PM_{2.5}$ mortality in long-term prospective cohort studies: cause–effect or statistical associations? *Environ. Health Perspect.* **106:** 535–549.

Gamble, J.F. and Lewis, R.J. (1996) Health and respirable particulate (PM_{10}) air pollution: a causal or statistical association. *Environ. Health Perspect.* **104:** 838–850.

Gilmour, P., Brown D.M., Beswick P.H., Benton E., MacNee W. and Donaldson K. (1997) Surface free radical activity of PM_{10} and ultrafine titanium dioxide: a unifying factor in their toxicity? *Ann. Occup. Hygiene* **41** (Suppl): 32–38.

International Commission on Radiological Protection. (1994) Human respiratory tract model for radiological protection. ICRP Publication 66. *Ann. ICRP* **24** (1–3): 1–482.

Kobzik, L. (1995) Lung macrophage uptake of unopsonized environmental particulates – role of scavenger-type receptors. *J. Immunol* **155:** 367–376.

Lambert, W.E., Samet, J.M. and Dockery, D.W. (1998) Community air pollution. In: *Environmental and Occupation Medicine*, 3rd Edn (ed. W.N. Rom). Lippinscott Raven Publishers, Philadelphia.

Lawther, P.J. (1961) Urban air pollution and its effects on man. In: *The Air We Breathe.* (eds S.M. Farber, R.H.L. Wilson). Charles C. Thomas, Springfield, USA, pp. 206–211.

Li, X.Y., Gilmour P.S., Donaldson K. and MacNee W. (1996) Free radical activity and pro-inflammatory effects of particulate air pollution (PM_{10}) *in vivo* and *in vitro Thorax* **51:** 1216–1222.

Lipfert, F.W. and Wyzga, R.E. (1995) Air pollution and mortality: issues and uncertainties. *J. Air Waste Manage Assoc.* **45:** 949–966.

Ministry of Health (1954) Mortality and morbidity during the London fog of December 1952. *Reports on Public Health and Medical Subjects No. 95.* HMSO, London.

Oberdörster, G., Ferin, J., Gelein, R., Soderholm, S.C. and Finkelstein, J. (1992) Role of alveolar macrophage in lung injury: studies with ultrafine particles. *Environ. Health Perspect.* **97:** 193–199.

Oberdörster, G., Gelein, R.M., Ferin, J. and Weiss, B. (1995) Associations of particulate air pollution and acute mortality: involvement of ultrafine particles? *Inhalation Toxicol.* **7:** 111–124.

Schrenk, H.H., Heimann, H., Clayton, G.D., Gafafer, W.M. and Wexler, H. (1949) Air pollution in Donora, Pa. Epidemiology of the unusual smog episode of October 1948. Preliminary Report. *Public Health Bulletin No. 306.* US Public Health Service, Washington DC.

Seaton, A., MacNee, W., Donaldson, K. and Godden, D. (1995) Particulate air pollution and acute health effects. *Lancet* **345:** 176–178.

Waller, R.E. (1991) Field investigations of air. In: *Oxford Textbook of Public Health*, 2nd Edn, Vol. 2 (eds W.W. Holland, R. Detels, G. Knox). Oxford University Press, Oxford, pp. 435–450.

Waller, R.E., Brooks A.G.F. and Cartwright J. (1963) An electron microscope study of particles in town air. *Int. J. Air Water Pollut.* **7:** 779.

Wilson, R. and Spengler, J. (1996) *Particles in Our Air: Concentrations and Health Effects.* Harvard University Press, Harvard School of Public Medicine, Boston, MA.

World Health Organization. (1987) Air Quality Guidelines for Europe 1987. *WHO Regional Publications, European Series No. 23.* WHO, Copenhagen.

World Health Organization. (1990) Acute Effects on Health of Smog Episodes. *WHO Regional Publications, European Series No. 43.* WHO, Copenhagen.

2 Airborne particles and their effects on health

Anthony Seaton

2.1 The lung and its defences

The birth of an air-breathing mammal presents it with a formidable problem. Lungs, previously filled with liquid, fill with air from which they must extract the oxygen essential to life. However, that air also contains many biological and other particles, any of which threaten the new life. Among these are bacteria, fungal spores and viruses, other life forms with which the mammal is in competition. The survival of the new animal depends upon evolutionary adaptations that have taken place in the lung to enable it to breathe without succumbing to infection.

At its simplest, the lung is a series of branching tubes of progressively smaller individual diameter but of increasing total diameter, leading to a thin alveolar membrane through which exchange of oxygen and carbon dioxide takes place. By the time human lung growth is complete at adolescence there are some 28 orders of division of the airways with about 25 000 terminal bronchioles leading to a total alveolar surface area of some 40–80 m^2. The vulnerability of this system to inhaled particles is illustrated by the importance of lung infections as a cause of death world-wide – tuberculosis and pneumonia are the main causes of death in the poor world and increasingly threaten health in the developed world. When the lung's defence systems go wrong, as in malnutrition, cystic fibrosis or AIDS, lung infection is the usual cause of death.

The lung's defences are both mechanical and chemical. The airways are lined by an epithelium, a self-regenerating system of cells, which forms a barrier to any inhaled organism and which secretes mucus with antibacterial properties. This liquid, familiar if produced in excess as sputum, lines the inner surface of the airways and is propelled from the small airways upwards to the throat by co-ordinated beating of the cilia

Particulate Matter, edited by R.L. Maynard and C.V. Howard

on specialized surface cells. Any inhaled organism deposited on to the airway wall is likely to be removed by the mucus and neutralized by the chemicals within it. Damage to this defensive system by, for example, cigarette smoke or inhaled toxic fumes will lead to inflammation and the symptoms of cough and sputum production known as bronchitis. Particles containing carcinogens such as polyaromatic hydrocarbons (PAHs), if deposited on damaged, poorly defended airways, may release the chemical where it can be metabolized within the airway, leading ultimately to lung cancer.

In the alveoli, the main lung defence is provided by macrophages, cells that are attracted to micro-organisms and ingest them. Macrophages have many functions, including the ability to digest some organisms, secretion of defensive chemicals and attraction of other inflammatory cells. Once they have ingested a particle, they transport it either up the airway or through the alveolar membrane into the lymphatic system that drains fluid from the peripheral lung to the lymph nodes at the lung root and thence into the blood stream. This system has evolved to ensure that virulent organisms are either removed or carried to places where the body's immune defences are concentrated and where the organism can be most effectively neutralized. One problem associated with the body's defence systems is that the chemical substances designed to kill invading organisms are themselves potentially toxic to the cells of the body itself. This process is usually self-limiting, but becomes a particular problem when the same mechanisms are switched on to deal with non-organic particles such as quartz or asbestos.

The effect of this system is to protect the body against inhaled organisms, unless a serious overload is inhaled. However, some organisms have evolved mechanisms to counter these defences and are therefore important causes of illness. For example, the tuberculosis bacillus is able to survive within macrophages and may live in them for years before causing disease, while the fungus, *Aspergillus fumigatus*, has developed mechanisms to resist being ingested by macrophages and therefore can in some circumstances survive and multiply in the lung and airways. The anthrax bacillus, a rare cause of disease nowadays, is transported to the lymph nodes where it causes a spreading and fatal infection around the heart. These adaptations are likely to be responses to evolutionary pressures in the soil, whereby the organisms have tipped the balance of survival in their favour. It is of interest that these same pressures have made soil micro-organisms an important source of anti-microbial drugs such as penicillin and streptomycin. Indeed, the search for a cure for tuberculosis was largely conducted by biologists who knew that somewhere in the soil an organism would be found that was able to counter the territorial aggression of *Mycobacterium tuberculosis*.

2.2 Biological particles

It seems likely that the defence mechanisms of the lung have evolved primarily to deal with biological particles, ranging in size from viruses

(20–300 nm) through bacteria (up to a few microns) to fungal spores which may be as large as 20 or 30 microns. These come in different shapes, from spherical to plate-like, acicular or fibrous, and their deposition in the lung is of course determined by the same physical factors as apply to non-biological particles. Their pathogenic effects are of fundamental importance to the survival of mankind as a species, and are responsible for all the great plagues of history including the current AIDS pandemic. Further discussion is outwith the scope of this chapter.

2.3 Mineral particles

2.3.1 The pneumoconioses

The relevance of mineral particles to man's health probably first became apparent in the Middle Ages with the recognition that work as a miner was followed by premature death from lung disease. The early Industrial Revolution led to a great increase in the numbers of people exposed to dust at work, and the condition known as silicosis was first described in the mid-18th century. It was observed that the pathological reaction was associated with very fine particles of quartz in the lung, and that such particles were capable of remaining suspended in the air for prolonged periods. Miners, stone grinders, tunnellers and foundry workers were shown to be at risk of silicosis, a disease that was characterized by nodular foci of dense fibrosis in the alveolar part of the lung, around the smallest terminal airways. These nodules characteristically grew and aggregated together to form large masses and extensive lung destruction, leading to failure of the right side of the heart and death.

In the first half of the 20th century it was recognized that a somewhat different form of what had by then acquired the generic name of pneumoconiosis, literally dust in the lung, afflicted coal miners, and this was given the name 'coal workers' pneumoconiosis' (cwp). This condition shares many features with silicosis but tends to be less progressive. It also has a greater tendency to be associated with the destruction of alveolar tissue known as emphysema. An interesting feature of cwp is that markedly different risks have been demonstrated in association with exposure to different types of coal, roughly in relation to its hardness or combustibility. It has been shown that the toxicity of coal varies depending on its mineral composition, and that its constituents may interact; for example, the silicates may reduce the toxicity of the quartz in the dust. This is in keeping with the concept that the surface activity of toxic minerals is important in their ability to cause disease.

The early 20th century also saw the first descriptions of diseases due to asbestos. At first this mineral was regarded as an occasional cause of a rare type of diffuse lung fibrosis, asbestosis, only afflicting people working with and exposed to high concentrations of it. It was then recognized that patients with this disease had a substantially increased risk of lung cancer, and in 1960 a further hazard, the malignant chest wall tumour, mesothelioma, was described in people exposed to asbestos. It became apparent that patients with mesothelioma had often been exposed to the mineral for rather short periods, sometimes only a few months. Compared to silicosis and cwp, the actual mass of inhaled asbestos necessary to cause serious disease appears trivial, and the threat to health of workers has therefore been quantified by counting fibres rather than by attempting to measure their mass. The need to do this arose from an understanding that it is the fibres of asbestos that are harmful and not any associated amorphous particles, and much research has been targeted at understanding how fibres cause disease. This work has suggested that four factors are largely responsible for determining their toxicity; their length, diameter, durability and surface properties. Length is important in that fibres longer than about 10 μm are not cleared efficiently by macrophages. Diameter is important in that fibres broader than about 3 μm are not able to penetrate deep into the lung. Durability is important in that fibres that dissolve in the lung, for example some glass fibres, do not persist long enough to cause disease. It is noteworthy that these are also characteristics that make fibres commercially desirable. Finally, some fibres of similar size and durability may have quite different toxicity, with respect for example to causing mesothelioma, and there is some evidence that this may be due to their different surface activities.

Studies of the adverse health effects of asbestos have given rise to understandable anxiety that other substitute fibres used in industry might have similar adverse effects on health. To date, however, this has not proved to be the case. Partly this may be due to better control over exposures to fibres of respirable diameter, and it would be unwise to dismiss the hazard posed by durable fine fibres in industry; all that can safely be said is that so far the risk appears to have been well controlled.

A simple overview of the mineral dust diseases is that the particles first need to be respirable, that is able to reach and be deposited in the alveolar part of the lung in sufficient amount. In general this means that as roughly spherical particles they should be below about 10 μm aerodynamic diameter, or as fibres less than about 3 μm in width. Non-fibrous particles will be removed by macrophages either up the airways or through the lymphatic system, and their ability to cause disease depends primarily on their toxicity to macrophages and other defensive cells. An overload of toxic particles will cause, first, damage and fibrosis of the lymph nodes, interfering with further removal, then destruction and fibrosis of lung tissue, characteristically around the smallest airways and along the lymphatic pathways. This tissue reaction is typically nodular, with normal lung in between. In contrast, the inhalation of fibrous particles provokes a reaction where the fibres lodge, around the airways, since these long fibres cannot be removed by macrophages, and this fibrotic lung reaction is diffuse rather than nodular.

2.3.2 Lung cancer

The chief cause of lung cancer is inhalation of cigarette smoke, and approximately one smoker in eight will die of this disease. Smoke contains many carcinogens, including particles carrying PAHs and radioactive elements such as polonium. In industrial situations, particles with adsorbed radon daughters are also a recognized cause of lung cancer in metal miners. The mineral particles to which coal miners are exposed, however, do not increase the risk of lung cancer. Asbestos fibres are undoubtedly carcinogenic, although there has been and remains much argument as to whether this is a direct effect of the fibres or is mediated indirectly by first causing fibrosis. Whatever the mechanism, exposure to sufficient asbestos to cause fibrosis results in a very substantial increase in lung cancer risk. The most durable of the asbestos fibres persist in the lung and as time passes migrate to its pleural lining where they may cause the tumour mesothelioma, characteristically 30–40 years after exposure first occurred. This disease has assumed considerable importance in the developed world because of its increasing incidence. Until 1970 it was regarded as a very rare condition, but data from the UK have shown a progressive rise, with over 1000 deaths occurring annually at present. Predictions based on knowledge of the amount of asbestos imported and the uses made of it suggest that these numbers will at least double over the next two decades before declining. In some trades, prolonged exposure to asbestos has entailed a 10% or greater risk of death from the disease.

There has also been much debate about the carcinogenicity of quartz. Several studies have shown that quartz-exposed workers have an increased risk of death from lung cancer, but the debate centres round whether the cancer has been caused by the quartz or by coincident exposure to cigarette smoke, radon or polycyclic hydrocarbons. This issue has not been resolved, but it seems that any increase in risk is very small compared to that from asbestos exposure. It is likely that different types of quartz, especially in terms of their surface activity and adsorbed elements, will differ in their ability to promote the development of cancer. The author's view is that exposure to sufficient quartz to cause silicosis increases the risk of lung cancer, with the corollary that measures to prevent silicosis will prevent this increase in risk.

2.4 Ambient air pollution

The concept of air pollution is taken to mean the addition to natural air of substances produced by human activity. Of course, natural air is far from a simple mixture of nitrogen, oxygen, carbon dioxide and some inert gases, and has always included not only the organisms discussed above but also particles and other gases generated by the natural activity of, for example, volcanoes and atmospheric photochemistry. It is likely that mankind's lungs have been exposed to relatively high concentrations of pollutants since man learnt to make fire and use it in caves. Studies in the 19th century in the UK showed that mortality rates increased in the winter time and that this was not simply an effect of temperature. In the mid-20th century it became apparent that urban smogs were associated with marked increases in mortality from heart and lung disease, and these observations led to smoke control legislation in the UK and Europe. In spite of subsequent and considerable reductions in the concentrations of particles in urban air, it has been recognized that increases in mortality and morbidity still occur in relation to rises in the level of particles in the air of major cities. The magnitude of such effects is now far smaller than in the 1950s, and recent effects in the UK have only been demonstrable with any confidence in the largest cities. However, the size of the effect on health in relation to unit rises in particle concentrations seems to be very similar wherever it has been studied world-wide, and there is a scientific consensus that quite small rises in concentrations of particles, even from baseline levels around 30 $\mu g\ m^{-3}$, will be associated with epidemiologically measurable increases in death rates and hospitalizations in urban populations of a million or so people.

These epidemiological associations have caused problems to toxicologists who wish to explain them in terms of cause and effect. It is difficult to conceive of such small doses of particles causing illness or death when even the most toxic particles, such as quartz, require to be inhaled in milligram doses over a prolonged period to cause serious harm. However, it seems likely that the effects are real and are not explained by some unknown confounding factor, so a biological explanation is required. One way out of this dilemma appears when we realize that the observed effects occur in large populations of people and not in the small groups of otherwise healthy rats studied by toxicologists. Such populations include the vulnerable elderly and sick, those close to death in whom only a small additional stress is sufficient to tip them over the edge. Moreover, a small overall rise in particle mass recorded at a background city site may represent greater concentrations closer to sources in that city. While individuals will all have a threshold of resistance to the potentially harmful effects of particles, populations may not. But even if this reasoning were correct, the toxicologist would like to have a mechanism by which urban particles could do sufficient harm to the lung of a vulnerable individual to cause illness and even death.

A further difficulty arises from the fact that urban particulate air pollution comprises largely carbon generated by combustion processes and ammonium sulphate and nitrate produced by atmospheric chemical reactions. None of these seems likely at first sight to be toxic, particularly in small doses. However, a clue has come from consideration of the size distribution of combustion particles and a realization that the smaller they are, the greater their surface area and therefore potential to carry toxic elements deep into the lung. The great majority of these particles are less than 1 μm in diameter and most fall within the size range 20–100 nm. It has been shown in experimental animals that such very small particles are capable of penetrating the alveolar epithelium very readily and causing lung inflammation at mass concentrations which cause no harm if less finely divided. There is now some evidence that this effect may be a consequence of the release of free radicals by a chemical reaction on their surfaces generated by the presence of adsorbed transition metals.

This possible mechanism of toxic lung damage from urban particles provides a means of explaining temporary worsening of disease in patients already suffering from severely damaged lungs. It is less easy to see why air pollution might cause people to have heart attacks, but at least two hypotheses have been proposed. The first suggested that lung inflammation can lead to release of chemicals from macrophages that cause changes in the clotting mechanisms of the blood. Unfortunately for those who formulated this hypothesis (including the author of this chapter), a direct test has failed to support it! However, the same study has shown that individuals' exposure to ambient particles relates inversely to the concentration of oxygen-carrying red cells in the blood in their veins – the more air pollution they are exposed to, the lower their red cell count. This has been interpreted as indicating that the red cells become stickier and get held up when they pass through the capillaries; it is possible that this impairs blood flow through the heart's own arterial blood supply sufficiently to cause heart attacks in the vulnerable. The second hypothesis has proposed that a neural reflex may be initiated by the irritant effects of air pollutants in the lung, leading to changes in heart rate and, by implication, increased liability to fatal changes in its rhythm. There is some support for this from studies of changes in people's heart rate in relation to exposure to particulate pollution.

2.5 The lessons

The story of the study of health effects of particles to date shows that an understanding of the epidemiological relationships between exposure and response can lead to effective preventive measures even in the absence of an understanding of the precise mechanisms involved. Furthermore, the call for further research before action is taken has in the past sometimes been used by industry to delay action. Effective action has been taken in most Western countries to reduce air pollution, with undoubtedly beneficial effects on public health in the large industrial cities. For the scientist now, the main challenge is to investigate the possibility of thresholds at population levels or to explain (perhaps in terms of vulnerable sub-sections of the population) why these are not demonstrable. It is plainly impossible to eliminate all harmful particles from the air, so it is desirable to know at what concentrations adverse effects are no longer demonstrable in large conurbations. If this is not possible, it is necessary to know how the vulnerable are best protected.

Several other intriguing questions remain to be answered with respect to particulate pollution. Of these, the most important is probably whether there are long-term cumulative effects as well as the short-term ones described above. Some evidence suggests that increased risks of early death from lung disease and lung cancer may be associated with living in polluted cities, and this is certainly biologically plausible. However, those who live in polluted places usually suffer multiple social disadvantages which also predispose to early death, making it difficult to be sure how much of the effect is due to pollution. No less intriguing is the observation that the epidemiological consequences of a rise in particle concentration seem to be much the same whatever the initial concentration, so that the population of a city with an average annual concentration of, say, 100 $\mu g\ m^{-3}$ will show no greater adverse effect to a rise of a further 20 $\mu g\ m^{-3}$ than will the population of a city with an average concentration of 20 $\mu g\ m^{-3}$. It seems that the absolute concentration of particles is not as important as the change of concentration. Whether one thinks of lung toxicity in terms of mass or number/surface area, this seems strange. Is the lung programmed to adapt to ambient levels of particles but to respond by inflammation to change in the level, and if so, why? What evolutionary pressures might have produced this behaviour?

Further reading

Further information on many of the points raised in this chapter will be found in:

Department of Health. Committee of the Medical Effects of Air Pollutants (1995) *Non-biological Particles and Health*. HMSO, London.

Dockery, D.W., Pope, C.A., Kanner, R.E., Villegas, G.M. and Schwartz, J. (1999) Daily changes in oxygen saturation and pulse rate associated with particulate air pollution and barometric pressure. *Res. Rep. Health Eff. Inst.* **83:** 1–28.

Doll, R. and Hill, A.B. (1956) Lung cancer and other causes of death in relation to smoking: a second report on the mortality of British doctors. *Br. Med. J.* **2:** 1071–1081.

Doll, R. and Hill, A.B. (1964) Mortality in relation to smoking: ten years' observations of British doctors. *Br. Med. J.* **1:** 1399–1410; 1460–1467.

Gilmour, P.S., Brown, D.M., Lindsay, T.G., Beswick, P.H., MacNee, W. and Donaldson, K. (1996) Adverse health effects of PM_{10} particles: involvement of iron in generation of hydroxyl radical. *Occup. Environ. Med.* **53:** 817–822.

Li, X.Y., Gilmour, P.S., Donaldson, K. and MacNee, W. (1996) Free radical activity and pro-inflammatory effects of particulate air pollution (PM_{10}) *in vivo* and *in vitro*. *Thorax* **51:** 1216–1222.

Ministry of Health (1954) Mortality and morbidity during the London fog of December 1952. *Reports on Public Health and Medical Subjects No. 95*. HMSO, London.

Morgan, W.K.C. and Seaton, A. (1995) *Occupational Lung Diseases*, 3rd Edn. W.B. Saunders, Philadelphia, PA.

Salvi, S., Blomberg, A., Rudell, B., Kelly, F., Sandström, T., Holgate, S.T. and Frew, A. (1999) Acute inflammatory responses in the airways and peripheral blood after short-term exposure to diesel exhaust in health human volunteers. *Am. J. Respir. Crit. Care Med.* **159:** 702–709.

Seaton, A., MacNee, W., Donaldson, K. and Godden, D. (1995) Particulate air pollution and acute health effects. *Lancet* **345:** 176–178.

Seaton, A., Seaton, D. and Leitch, A.G., (eds) (1989) *Crofton and Douglas's Respiratory Diseases*, 4th Edn. Blackwell Scientific Publications, Oxford (5th edition to be published in 2000).

Seaton, A., Soutar, A., Crawford, V., Elton, R., McNerlan, S., Cherrie, J., Watt, M., Agius, R. and Stout, R. (1999) Particulate air pollution and the blood. *Thorax* (in press).

West, J.B. (1995) *Respiratory Physiology–The Essentials*, 5th Edn. Williams and Wilkins Co., Baltimore, MD.

3 Microscopy and the characterization of particles

Frederick D. Pooley and Milagros Gálvez de Mille

3.1 Introduction

The chemical and physical characteristics of airborne particulate matter in urban environments are very complex and dynamic. They are a function of the many sources that contribute to its formation and of the variable weather conditions that may modify its composition. The characteristics of ambient particulate material may change significantly with time as the emissions from local sources vary in their magnitude and identity. The import of foreign airborne material by atmospheric transport to any particular locality may also frequently serve as a chemical base to which locally generated material will be added to form a unique composition at any one time.

Because of the complex composition of airborne particulate matter its analysis can only be attempted by employing sequential analytical methods and a variety of analytical techniques. Having collected as complete a chemical and physical description of the properties of a sample as possible these can then be attributed to its individual component phases and related to particular sources.

The importance of the quantification and identification of material of non-biological origin that may be found in ambient air has received increasing attention in recent years. This is related to the demonstrable association of human mortality, morbidity and environmental impact with ambient levels of particles. It has become apparent that a more permanent and extended examination of the materials implicated as possible causes of these effects is required. No major effort in the research into the environmental influence of particulate aerosols can proceed without adequate information concerning the quality and quantity of the individual particulate materials involved. Such information contributes

Particulate Matter, edited by R.L. Maynard and C.V. Howard

significantly to the design and scientific interpretation of biological experimentation and epidemiological studies.

The measurement and determination of the chemical and physical characteristics of particulate aerosols is essential in the application of environmental control by identifying sources and describing the origin and fate of the various products of atmospheric chemical interactions. One of the most useful techniques employed in such investigations is that of analytical transmission electron microscopy.

3.2 Sampling particulate aerosols for analysis

The analysis of airborne particulate samples is initially very dependent upon the accuracy and manner with which the material is sampled from the air. Care should always be taken in sampling to ensure that the property to be measured is not altered in any way by the techniques employed to collect, store or transfer the sample to be analyzed.

Methods of sampling can be generally divided into bulk or individual particle collection methods. The former technique will be useful in providing information on the airborne mass concentration and the bulk chemical and physical characteristics of the aerosol. It is impossible to re-create the true physical and chemical characteristics of particulate aerosols from bulk samples, as both its collection and redispersion will dramatically alter its properties.

The analysis of bulk particulate samples can provide basic chemical compositional data but limited morphological information. The need to disperse particles in a fluid prior to the examination of individual particles will result in the loss of important chemical and physical information by dissolution of components and disaggregation of agglomerates.

Sampling particulate aerosols on a size selective aerodynamic diameter basis enables size-differentiated compositional characteristics to be established. Special collection and careful preparation of size fractions allows the morphology and chemistry of particles to be observed in tandem when analytical electron microscopic techniques are employed. This information is important for describing the hazards associated with such material. It also provides accurate physical information concerning the aerosol, enabling its concentration on a mass and particle number basis to be determined. This also allows estimation of that fraction of the aerosol which is respirable for the purpose of risk assessment.

It is to be expected that the various components which combine to form airborne particulate samples will vary in their individual size characteristics. Particles of a secondary origin which form by nucleation and subsequent chemical growth will possess size distributions ranging from the finest to the most coarse particles. Particles which owe their origin to mechanical reduction of materials and natural lithological processes will, in the main, possess coarser size characteristics.

When collecting bulk samples of particulate aerosols it is essential to select suitable filter substrates which do not interfere with subsequent analysis of the sample. Glass fibre and other loose fibre filter constructions can contaminate samples when they are removed from the filters for analysis, thus limiting the data collected from them.

Bulk filter samples will consist of a mixed deposit comprising insoluble and soluble solid particulate materials and fluid droplets. The individual identity and particulate characteristics of the soluble solid and fluid droplets will be lost in the subsequent resuspension of the bulk aerosol sample with a suitable fluid media. General chemical information can be collected by analysis of the media to describe the soluble components of the aerosol sample, but the preparation procedure will result in a loss of information relating to fragile and more volatile fractions of the aerosol.

Bulk samples are most useful for comparison of the composition of airborne material collected at different times and in different locations. They are not limited in size for analytical purposes as the quantity collected is simply a function of the amount of air sampled. Total particulate samples can be collected on simple open filter heads whilst size selective sampling, employing an elutriator prior to the collection filter, can be used to collect fractions of airborne material with aerodynamic sizes of less than a specified size, such as PM_{10} and $PM_{2.5}$. These size fractions are defined as particles that pass through a size selective inlet with a 50% efficiency cut-off at 10 or 2.5 μm aerodynamic diameter.

Bulk samples of these particular aerosol size fractions are used extensively and world-wide to define the most relevant biologically active fractions of airborne material (Burton *et al.*, 1996; Chih-Shan Li, 1994; Chow *et al.*, 1994; Watts *et al.*, 1988; Whei-May *et al.*, 1994).

Size selective sampling of particulate aerosols has many advantages over bulk methods. It allows samples to be collected in such a form that both the physical and chemical state of the material is essentially preserved. Particle size fractions can be collected and examined to establish both individual physical and chemical characteristics. The cascade impactor is one such device which retains particles at various stages as discrete aerodynamically size-fractionated samples. These can then be compared by a variety of techniques, including microscopic methods. The mass distribution and quantification of soluble and insoluble species between the size fractionated samples can also be investigated using this sampling method and the results combined to produce a total particulate analysis.

Size selective sampling methods can be controlled to produce samples suitable for the application of a variety of analytical techniques. For optical and electron microscopy, samples with individual particles well dispersed are required. These can be collected over very short time periods onto substrates which can be examined directly with no further preparation of the samples. Although volatile substances such as water may be lost from the particles that are collected, the remaining residues are sufficient to identify the nature of the particles and their chemical composition.

3.3 Optical microscopic techniques

Optical microscopic techniques have been and are still the most inexpensive and practical methods employed to examine aerosol particles. The optical examination of mineral particles is a well-established mineralogical procedure (Kerr, 1977) but it is normally confined to the study of particles which are very much larger than those encountered in aerosol samples. These techniques are mainly applicable to particles which are translucent, as very fine opaque particles are extremely difficult to examine by normal optical methods. To identify particles by optical means requires that certain optical properties of the unknown materials have to be established and compared with reference information. The process involves a sequence of procedures which include the description of colour, form, indices of refraction, and isotropic properties. Identification will then finally depend upon the possibilities suggested by each of the properties examined.

Frequently, identification will be uncertain, and materials of similar optical properties will be confused. It is only the larger translucent crystalline particles in a dust specimen that can be readily characterized by their optical properties, a task which will be made even more difficult when several mineral phases are present, as each must be considered independently.

There are certain inadequacies in the use of optical microscopy for the examination of airborne dust specimens. The most obvious is the limitation of resolution of the optical microscope itself. It is claimed that particles as small as 0.2 μm can be observed by optical methods, but it is recognized that no measurement of optical properties can be made from particles which are so small (McCrone, 1978).

Optical microscopy is most useful in studying the larger particles that occur in dust specimens, and in this role can only be considered to be useful for obtaining morphological information. It can provide data concerning the number, concentration, size and shape of optically visible particles which may be of great value for the control of atmospheric pollution that presents a nuisance rather than a hazard to health.

Optical microscopic techniques continue to extend with developments such as dispersion staining (McCrone and McCrone, 1976) for the determination of specific particulate species. These are again only useful in those situations where particles are collected from a known source and which have been previously characterized.

3.4 Electron optical techniques

Early applications of electron optical equipment, in research on airborne particles, started in the early 1950s when the first transmission electron microscopes (TEMs) were employed to produce a more accurate description of the size and shape of dust particles (Cartwright *et al.*, 1956). The TEM was initially limited in its analytical capabilities to a morphological examination of dust material, together with the ability to produce electron diffraction patterns from crystalline phases, providing information about the structure and possible identity of certain particles in a mixture. As the TEM was developing, the production of equipment using finely focused electron beams in a scanning and fixed mode to produce electron optical images of surfaces and also to generate X-rays characteristic of the elements contained within the irradiated particle surfaces, was occurring.

The analysis of such X-rays would yield quantitative chemical information and provide a microchemical analysis capability of great importance in aerosol particle characterization. Modern electron optical equipment now incorporates both transmission and scanning facilities together with microanalysis facilities which can be shown to have evolved over a number of years (Pooley, 1981).

The basis for the analysis of particulate aerosols with electron optical equipment has been described (Pooley, 1977), and methods for the determination of size, shape, number, composition, identity and quantity of the various components of a particulate mixture have been specified. The analytical transmission electron microscope (A.T.E.M.) has been shown to be capable of detailed analysis of complex solids (Pooley, 1975; Rowse *et al.*, 1974) and liquid based droplets (Morgan *et al.*, 1975).

It has been demonstrated that with suitable electron optical equipment it is possible to analyze a dust sample to obtain morphological information, i.e. size and shape of particles, and also accurate chemical information from both single and large numbers of particles. The chemical information from single particles can serve as the basis for an identity of the materials contained in an aerosol sample, while the analysis of large numbers of particles provides a semi-quantitative chemical analysis of the whole aerosol. Employing chemical data collected from both single particles and bulk samples, it is possible to estimate the percentage composition of each chemical phase in a sample. These are normally results which are not possible to obtain by other analytical techniques.

Electron optical techniques are not limited by sample size as they are based on the ability to analyze single dust particles and they are unique in their ability to characterize particulate aerosol samples provided they have been correctly sampled and prepared for analysis.

3.5 Bulk analytical techniques

An approach to the characterization of bulk sampled particulate aerosols consists of the subdivision of the samples into four fractions for the purpose of analysis. The fractions sub-divide samples into convenient parts which are useful for describing the samples on the basis of their secondary and primary components and relate closely to the sources from which they are likely to have originated.

These fractionations are based on the following:

(a) the aqueous soluble material;
(b) material soluble in organic solvents;
(c) elemental carbon based material;
(d) the insoluble inorganic content of samples.

Water-soluble material can be very adequately described by a variety of wet chemical techniques which include ion chromatography, atomic absorption spectroscopy and inductively coupled plasma emission spectroscopy amongst others. Water-soluble carbon-based material can be quantified employing aqueous carbon analysis. The usefulness of these techniques is closely related to the quantity of airborne material collected and the sensitivity of the techniques employed. Their application entails sampling large quantities of air to obtain adequate material to perform an accurate analysis. The mass of aqueous dissolved material can be estimated from the concentration of species and the volume of aqueous solution in which they were collected. This value can be compared with the weight loss of treated samples for an independent comparison.

The insoluble elemental carbon content of samples can be determined by analysis of particulate residues after the removal of water-soluble and organic-soluble material. If sufficient insoluble material is available it can be determined by treatment of a sample portion by carbon analysis methods in which the carbon content is determined by the generation of carbon dioxide in an oxygen atmosphere at high temperature. This technique is very destructive, however, and the residues remaining are not suitable for further investigation.

Another technique which provides similar results consists of the treatment of the insoluble residue by ashing in an oxygen plasma at low temperature between 100 and 150°C to remove elemental carbonaceous material and determination of the sample weight loss. The residue of samples after such a treatment provide the insoluble inorganic fraction of the bulk aerosol sample which can be examined by a variety of techniques, including X-ray fluorescence, X-ray diffraction, optical and analytical electron microscopic techniques. These techniques are quantitative or semi-quantitative in nature and can be employed to provide a very detailed description of the refractory component of a particulate aerosol sample.

The sequence in which these various methods can be employed has been described (Pooley and Mille, 1999). They can produce a very detailed description of an aerosol sample, but the data obtained cannot be adequately recombined to describe the individual particulate components of the aerosol sample. Similar procedures can be utilized to analyze the size fraction of a particulate aerosol collected in the form of aerodynamically sized fractions by cascade impactor type instruments (Airborne Particulates Expert Group, 1999). This provides a more detailed description of the bulk chemistry of the aerosol samples and the size distribution of specific chemical components allowing the composition of the biologically relevant PM_{10} and $PM_{2.5}$ fractions to be compared.

3.6 Characteristics of particulate aerosols

Because of the complexity of urban aerosols it is convenient to discuss their characteristics on the basis of the subdivisions previously outlined as aqueous soluble, organic soluble, elemental carbonaceous, and insoluble inorganic. The proportions of these fractions in any particular sample will be dependent upon geographical location, proximity to industry, population and weather conditions. At any particular location the concentration of these general components in the atmosphere may vary considerably during the day corresponding to cyclic industrial activity and mobility of the resident population. They will be mainly anthropogenic in origin but contributions from natural sources will often be significant, being dependent on global scale activity in the atmosphere and the creation of aerosols due to local weather conditions and proximity to the sea.

Aerosol particulates are continuously generated at various rates due to atmospheric chemical reactions during the day and night, and given favourable atmospheric conditions when air movement and dry conditions prevail, can increase dramatically in concentration to allow considerable increases in specific aerosol components.

3.6.1 Aqueous soluble aerosols

The water-soluble component of particulate aerosols in urban air always represents a significant proportion of the material sampled and may average between 25 and 35% by mass. It can rise dramatically during pollution episodes and represent the majority of the suspended particulate material which may be sampled. Most of the water-soluble particulate aerosol component is classified as secondary particulate matter. This material owes its origin to atmospheric chemical reactions of gaseous species which result in the nucleation of new particles and also the attachment of reaction products to pre-existing particles.

Soluble inorganic particulate species can be divided into chemical groups which may include the following materials:

(a) Acid aerosol particles which are mainly formed by the photochemical oxidation of sulphur dioxide by free radicals in the atmosphere. The sulphuric acid vapour formed may nucleate to form sulphuric acid droplets and grow by incorporation of water vapour and coalescence with each other by Brownian aggregation. This acidity may be neutralized or partially neutralized by coagulation with particulate basic species or more likely the adsorption of gaseous ammonia from the atmosphere to form a range of secondary particles in which ammonium sulphate may predominate. These reactions take place on a timescale of hours, and are controlled by the level of active radicals produced by sunlight, sulphur dioxide and ammonia levels in the atmosphere. Similar pathways exist for the formation of nitric acid and subsequent nitrate reaction products by the oxidation of the nitrogen oxide gases. These chemical speciations and transformation reactions have been extensively studied. They are also the subject of extensive monitoring exercises in a number of industrialized countries (Harrison and Sturges, 1984; Review Group on Acid Rain, 1997).

(b) Sulphate is the most common species found in the soluble component of particulate aerosols and may exist in combination with ammonium, sodium and potassium ions. Because the majority of the chemical analyses of particulate aerosols have been obtained from bulk samples, it is not clear whether the sulphate salts exist as separate particulates or as combinations with other elemental material. Most of the particulate sulphate in urban air owes its origin to the oxidation of gaseous sulphur dioxide, but sulphate of a marine origin in the sodium and potassium form does contribute to the quantities measured. No information exists which indicates the proportion of water soluble sulphate which may be associated with insoluble particulate material in urban aerosol samples.

(c) Soluble nitrate like the sulphate fraction, is formed by the atmospheric oxidation of nitrogen oxides, but mainly nitrogen dioxide. Its concentration in urban particulate aerosols is closely linked to combustion processes, both transport and industry related. Nitrate and its associated reaction species are known to vary relatively rapidly on a diurnal basis due to a significant dependence upon the photochemical production of active radicals in the atmosphere. The atmospheric production of ozone which can be related to photochemical atmospheric reactions of organic vapour pollution is often a significant part of the oxidation reaction sequence of nitrogen oxides. Nitrates are assumed to be present in particulate aerosol samples in chemical association with ammonia, and are considered to exist associated with pre-existing particles with which they nucleate, and not as separate particles. One such association is with sea-salt particles which results in the production of sodium and potassium salts and a subsequent release of hydrogen chloride gas.

(d) The soluble chloride content of aerosol particulate material is mainly derived from the sea. The exceptions to this rule are samples collected in the vicinity of industrial activity where there may be a considerable emission of hydrogen chloride. Concentrations of chloride in the air of coastal regions can be considerable when weather conditions allow high onshore winds to prevail as they will suspend large quantities of sea water and associated salts. Chloride in the form of halide crystals from these marine sources occur almost exclusively in the coarser size fractions of coastal urban aerosol samples. Chloride ions are also considered to be associated with ammonia as a direct result of reactions of the latter material with hydrochloric acid emissions by industrial pollutant sources and also as a consequence of the release of hydrochloric acid by nitric acid displacement of chloride in sea-salt particles.

(e) The cations most commonly measured in the aqueous soluble fraction of particulate aerosols are ammonium, sodium, potassium, calcium and magnesium.

A considerable proportion of these species is derived from natural sources, but industrial and transport sources may also be important. Particulate ammonium has been the subject of a number of studies (Review Group on Acid Rain, 1997) in which it has been noted that reduced nitrogen species exist as both gaseous and particulate phases, the majority being in the latter form, and their concentrations have been mapped on a European scale (Lee *et al.*, 1997).

Sea-salt particles are a major source of sodium and a major source of other cations. Few detailed analyses of the aqueous soluble fraction of urban aerosol samples are available, but studies (Airborne Particulates Expert Group, 1999) have shown that iron, cobalt, nickel, zinc, amongst others, can be detected in the soluble fraction from bulk aerosol samples. Whether these are as a result of atmospheric reactions between acidic aerosol species or a consequence of similar reactions as a result of particulate collection procedures is not known.

Results from the examination of the water soluble component of bulk samples of aerosol particulates are unable to provide detailed characteristics of this soluble phase when it is airborne. However, analytical transmission electron microscopic examination of size-selective particulate samples has shown that information concerning the physical and chemical characteristics of this material can be collected.

An illustration of a particulate aerosol size fraction collected by size selective sampling in less than 1 h is shown in *Figure 3.1*. Short duration sampling is required for the study of individual particles to ensure they are adequately spaced for examination, with no overlap or coalescence occurring and to preserve if possible their original physical and chemical characteristics.

The electron micrograph illustrated in *Figure 3.1* contains images of a particulate aerosol sized fraction representing material with an aerodynamic size range between 0.43 μm and 0.65 μm. A number of the particles are obviously solid in appearance, but many possess the appearance of desiccated fluid droplets with dried material, sometimes crystalline in appearance, within the particulate deposit. The images suggest that many of the particles observed were originally fluid in

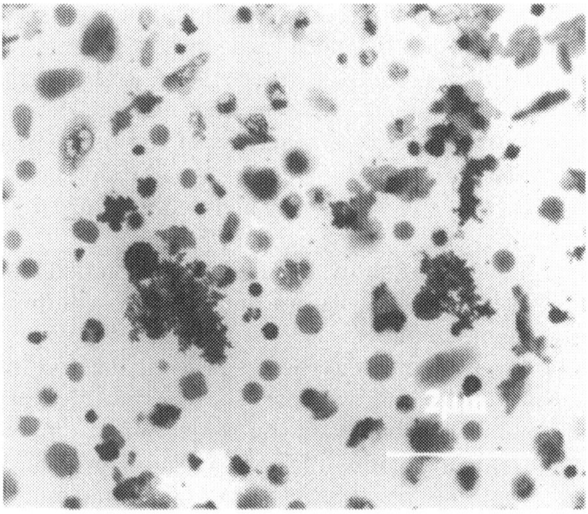

Figure 3.1. A >0.43–<0.65 μm aerodynamic diameter size fraction sample from an urban aerosol sample collected in 1 hour.

nature: the images observed represent the desiccated dissolved material they contained. A semi-quantitative chemical analysis of these individual deposits show that they have a variable chemistry. The average chemical composition of water soluble particulate aerosol size fractions, which include these droplets, has been found to change with increase in aerodynamic particle size.

These chemical differences are illustrated by *Figure 3.2 (a)* and *(b)* in which the change in average composition of bulk individual fractions of an urban aerosol sample are compared on the basis of aerodynamically sized fractions. It can be seen from *Figure 3.2(a)* that the anion content displays a general trend of sulphate reduction with particle size and an increase in chloride levels, while the nitrate content of size fractions are comparable. There are similar trends in the variation of the cation content of the size fractions. Ammonium contents generally decrease with size, while sodium increases before dropping dramatically in the largest size fraction. There is also a general increase in soluble magnesium and calcium with particulate size, while potassium levels are similar throughout the range.

The average analysis of single droplet-like fluid particles tend to confirm the chemical trends observed with the analysis of bulk size fractioned samples (Airborne Particulates Expert Group, 1999). They also reveal the presence of other chemical constituents of these particles, some of which may contain very small associated insoluble components. An illustration of results comparing these changes in the average semi-quantified fluid droplet composition is given by *Table 3.1*, in which contents of iron, silica and other elements were measured in some droplets. It is possible that these other components of the fluid droplet-

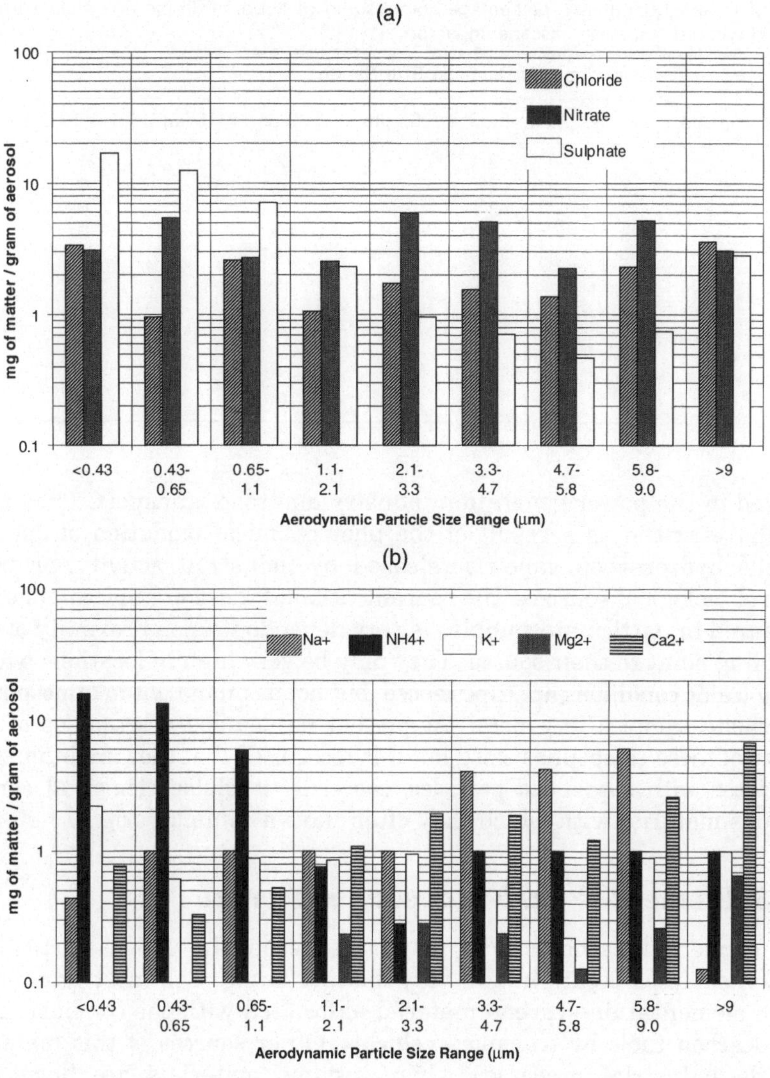

Figure 3.2. The change in composition of water soluble anions (a), and cations (b), in aerodynamically sized fractionated samples of an urban aerosol sample.

like particles may be related to the original solid condensation nuclei around which the fluid droplet may have formed.

3.6.2 Organic soluble aerosol material

The organic solvent soluble material present in particulate aerosols consists of a complex mixture of organic compounds. These materials will have arisen mainly as a result of the release into the atmosphere of volatile organic vapours or partially combusted hydrocarbons from fuels

Table 3.1. Semiquantitative percentage composition of airborne droplet-like fluid particles sampled with an Andersen cascade impactor

	Aerodynamic Diameter Size Range			
Elemental oxide	<0.43 μm	0.43- <0.65 μm	0.65- <1.1 μm	1.1 <2.1 μm
Na_2O	3	2	2	21
MgO	2	2	0	3
Al_2O_3	2	3	3	0
SiO_2	5	8	10	0
SO_4	59	73	66	13
ClO	5	4	3	47
K_2O	9	5	5	3
CaO	3	0	1	0
MnO	0	0	3	0
FeO	11	2	6	13
ZnO	0	1	0	0

utilized in the power generation industry and road transport. They may also have arisen as a result of the photochemical oxidation of specific volatile hydrocarbon vapours released by industrial activity, or from natural biological sources. The concentration of these organic compounds measured in particulate samples is very dependent upon proximity of the sampling point to their source. They may be very high in locations where heavy traffic conditions are experienced, but not as prominent in atmospheric particulate samples in general. At present no detailed descriptions of the nature of these compounds exist, but it is recognised that they are intimately associated with suspended particles, especially insoluble elemental carbonaceous material with which they often have a common origin.

3.6.3 Elemental carbonaceous material

The elemental carbonaceous material present in urban particulate aerosol samples occurs in two very distinct forms. The most prominent is the elemental fine carbon material associated with the combustion of hydrocarbon fuels by transport vehicles. Other sources of this material include industrial processes where organic materials are burnt for generating power or for waste disposal purposes. The second form of carbon based material occurs as carbonates which arise from natural sources such as limestone rocks or as particulate dusts generated by the construction and manufacturing industries, where products such as Portland cement and other carbonate mineral products are employed.

Elemental carbon-based particles generated from traffic are very distinct in appearance. They consist of chain-like aggregates which are difficult to define in terms of any specific dimension. The physical appearance of this elemental carbon material is illustrated by *Figure 3.3* which is an electron micrograph of a size selective sample with an aerodynamic diameter <0.43 μm. It can be seen that the small aggregate particles from which this material is formed vary in size and this may be

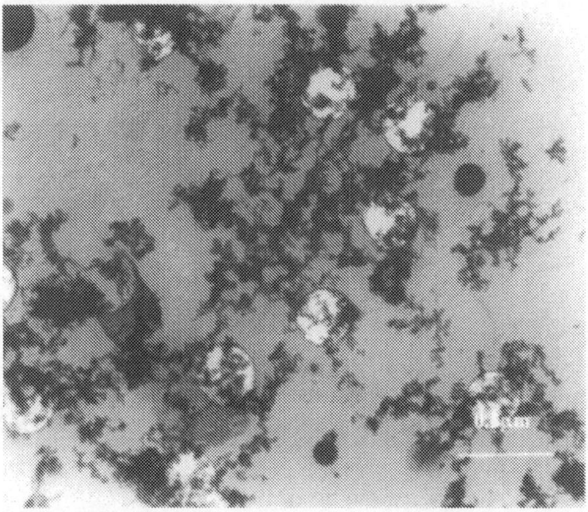

Figure 3.3. Elemental carbon-based particles detected in the <0.43 μm aerodynamic size selected fraction of an urban particulate aerosol sample prepared to direct transference to preserve morphological characteristics.

related to their origin. Material produced from the combustion of petrol, diesel or heavy fuel oil has been observed to generate this material with varying physical characteristics. Examination of size fractionated aerosol samples has shown that this elemental carbon material is concentrated mainly in the finer aerodynamically sized fractions. When it is observed in coarser fractions it consists of large agglomerates of the same material. It is rarely observed in the coarser particle size fractions of airborne dust. Chemical analysis of this material has shown it to possess a variable minor inorganic chemical content, including such elements as silicon and a range of heavy metals. The variable elemental content originates from the trace elemental content of the original hydrocarbon fuel from which carbon based particulate matter were formed.

The insoluble carbonaceous material in the form of carbonates is found mainly in the coarser size fractions of airborne dust. They occur mainly in the form of calcium and magnesium carbonates as the mineral forms of calcite and dolomite. The form and content of carbonate material in atmospheric aerosols has been examined by Clarke and Karani (1992) who confirm the concentration of these materials in the coarser sizes of particulate aerosols. The distribution of carbonaceous material between the aerodynamic size fractions of an urban aerosol sample and a comparison of its mass contribution to the overall composition of the dust is illustrated by *Figure 3.4*. In this illustration the relative amounts of the water soluble, insoluble carbonaceous and insoluble-inorganic components of an aerosol sample are represented. It can be seen that the finest fraction contains a significant amount of insoluble carbonaceous material and this is almost entirely in the form of elemental carbon. The

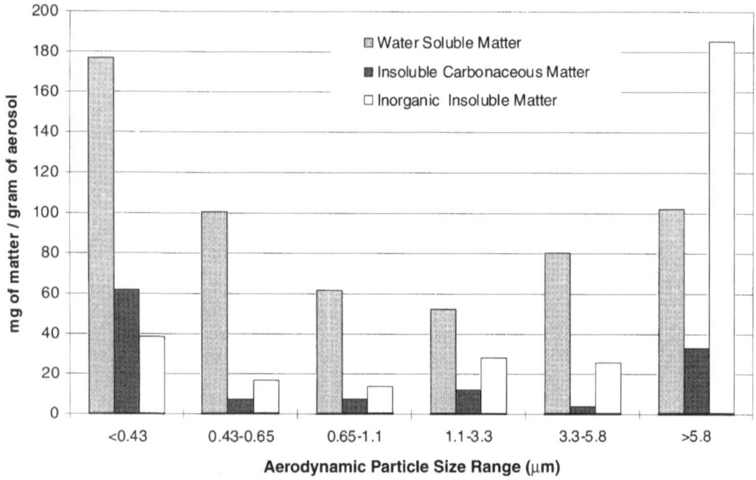

Figure 3.4. Distribution of insoluble carbonaceous water soluble and insoluble inorganic material between aerodynamic size fractions of an urban aerosol sample.

levels of carbonaceous material fall as particle size increases, but the level of this material in coarser fractions increases and is represented mainly by the presence of carbonate material.

3.6.4 Insoluble inorganic aerosols

The insoluble inorganic compounds of particulate aerosols are always a complex mixture of materials. They are generated from a range of natural and anthropogenic sources, any of which may be predominant in its contribution to this fraction of airborne particulate material. Insoluble inorganic material is present in all size fractions of particulate aerosols, but is a major component of the coarser size fractions, as illustrated by *Table 3.2* and also by *Figure 3.4*. When this material is industrial in

Table 3.2. Percentage mass distribution of components in aerodynamic size fractions of an urban particulate aerosol sample

Aerosol aerodynamic particle size range (μm)	Percentage aerosol component			Total
	Water soluble matter	Insoluble carbonaceous matter	Water insoluble matter	
<0.43	17.7	6.1	3.8	27.6
0.43–0.65	10.0	0.7	1.6	12.3
0.65–1.1	6.1	0.7	1.3	8.1
1.1–3.3	5.2	1.2	2.8	9.2
3.3–5.8	8.0	0.3	2.5	10.8
>5.8	10.1	3.2	18.5	31.8
Total	57.1	12.3	30.6	100

origin it may represent a significant proportion of the finer airborne particulate material.

Natural contributions of inorganic material to aerosol samples are more concentrated and detectable in the coarser size fractions. Finer inorganic material of an industrial origin consists mainly of metallic particles with a small contribution of silicate rich fly ash from fossil fuel combustion together with only a minor contribution of particles from natural sources. Coarser particles are predominantly natural in origin, consisting mainly of a mixture of silicate particles and naturally occurring oxide minerals of iron and titanium. *Figure 3.5* illustrates the appearance of an insoluble fraction of an aerosol sample collected in the vicinity of a large steelworks, the majority of the particles illustrated being inorganic materials.

Many of the particles observed in *Figure 3.5* are of an industrial origin, as illustrated by their spherical shape. Their chemical composition was found to range from very iron rich material to combinations of iron, manganese and zinc, with trace levels of silica, nickel and chromium. A comparison of the change in chemical composition with variation in aerodynamic particle size for this particular insoluble inorganic aerosol component is shown in *Table 3.3*. The size fractionated material was found to contain a range of elemental components but the chemistry of each fraction was dominated by the presence of iron oxide and silica. Distinct trends are observable in the data, with iron and other heavy metals decreasing in proportion with an increase in particle size and the silica content, aluminium and the alkali metals increasing in proportion. With careful characterization of particulate emissions from industrial sources it is possible to identify their presence and mass contribution to the insoluble inorganic composition of aerosol samples.

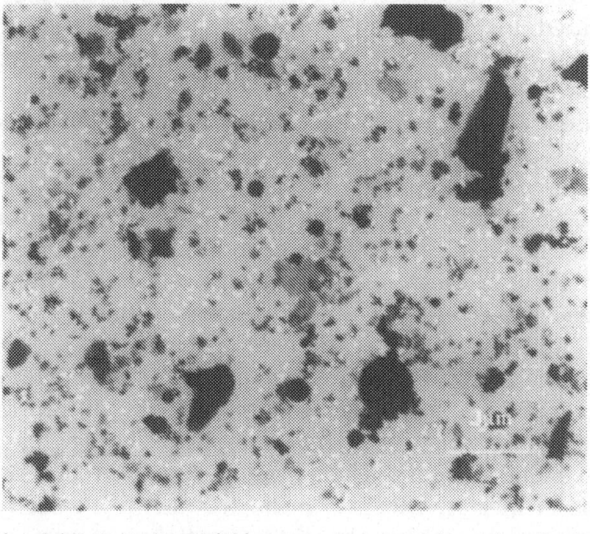

Figure 3.5. Insoluble aerosol material from an urban aerosol sample collected in the vicinity of a large steelmaking complex.

Table 3.3. Percentage semiquantitative oxide composition of the <1 μm, 1–<2.5 μm, >2.5 μm aerodynamic diameter fractions obtained from the insoluble inorganic component of the aerosol sample illustrated by *Figure 3.5*

Elemental oxide	Aerodynamic size fraction		
	<1 μm	1–<2.5 μm	2.5 μm
	Proportion of insoluble inorganic aerosol content		
	33	17	50
	% Elemental oxide content		
Na_2O	1	2	3
MgO	2	2	2
Al_2O_3	2	5	9
SiO_2	15	34	57
P_2O_5	1	nd	nd
SO_3	3	4	2
Cl	1	<1	1
K_2O	1	2	2
CaO	1	2	2
TiO_2	<1	1	nd
V_2O_5	nd	<1	<1
CrO_3	1	nd	<1
MnO	2	1	nd
FeO	68	43	21
CoO	nd	<1	nd
ZnO	3	1	1
Pb	1	1	nd

nd, not detected.

The contribution of minerals from natural sources to the coarse particle fraction of aerosol samples is often considerable. These sources may be local and related to the activity of surface mineral industries, the intensity of agricultural activity, or climatic conditions producing arid, sparsely vegetated soil cover. If this is the case then the composition of the inorganic dust will be predictable from an examination of the mineral content of local rocks and soils.

A further source of inorganic aerosol particles is the aeolian dust which may be generated on a very large scale by air turbulence in the more arid areas of the world, or by intense volcanic activity. Particles from these sources may be transported world-wide by large-scale volumetric movements of air in the atmosphere. This imported material can, under certain weather conditions, make a significant contribution to the inorganic aerosol component. *Figure 3.6 (a)* and *(b)* illustrate the appearance of this aeolian material. It represents the fine and coarse components of a dust sample which originated in the vicinity of the Sahara desert. The larger particles in this sample are rounded grains of quartz, feldspar, mica, together with natural oxides of iron and titanium, while the finer particles consist mainly of clay minerals.

(a) (b)

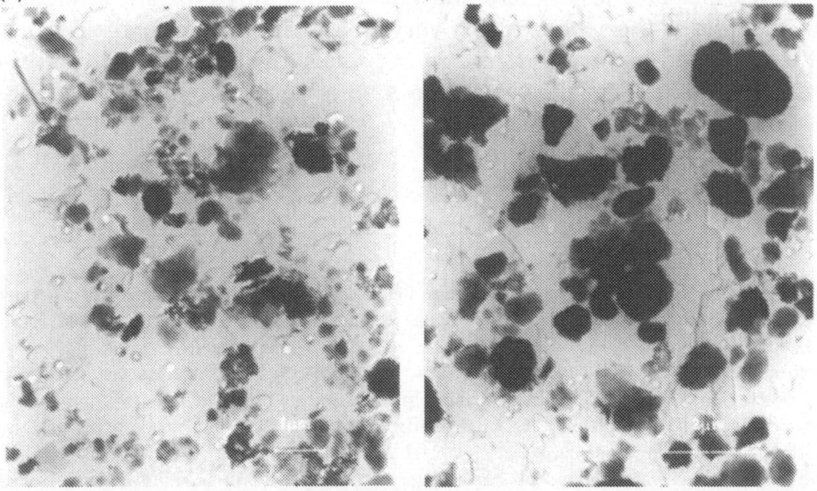

Figure 3.6. Fine and coarse inorganic particulates prepared from an aeolian dust sample which originated in the region of the Sahara desert

3.7 Conclusions

Our knowledge of particulate aerosol samples is not complete. Information concerning both quality and quantity of the organic solvent soluble components of samples and the distribution of aqueous soluble components between solid and liquid phases have still to be determined, together with their possible variation with particle size.

It is only possible to obtain the complete characterization of particulate aerosol samples by employing a range of analytical methods and techniques. Of initial importance is the selection of the aerosol sampling methods and the treatment of material collected to preserve in as much detail as possible the quality of the particles to be measured. Chemical and physical information collected from bulk aerosol samples, which can be divided into soluble and insoluble fractions, can be combined to provide a detailed description of the airborne material which can be employed to identify its origin or hazardous nature. Providing a better description of the biological potential of particulate aerosol samples and assessing specific characteristics of aerosol particles, requires that sampling techniques which fractionate material on an aerodynamic basis and preserve the physical and chemical characteristics of individual particles are used. Advanced analytical electron microscopical techniques can then be applied to provide detailed information concerning the size distribution, morphology and chemical composition of both solid and fluid particulates.

Application of these micro-analytical procedures is an interesting development as they require only very small samples for analysis. These can be collected in very short sampling times and offer the possibility of assessing rapid changes in the composition of particulate aerosol samples.

References

Airborne Particulates Expert Group (1999) *Source Apportionment of Airborne Particulate Matter in the United Kingdom.* Report of the Airborne Particulates Expert Group, January.

Burton, R.M., Sur, H.H. and Koutrakis, P. (1996) Spatial variation in particulate concentration within Metropolitan Philadelphia. *Environ. Sci. Technol.*, **30 2**: 400–407.

Cartwright, J., Nagelschmidt, G. and Skidmore, J.W. (1956) The study of air pollution with the electron microscope. *Q. J. Roy. Met. Soc.*, **82**: 82.

Chih-Shan Li. (1994) Elemental composition of residential indoor PM_{10} in the urban atmosphere of Taipei. *Atmos. Environ.*, **28 (19)**, 3139–3144.

Chow, J.C., Watson, J.G., Fujita, E.M., Lu, Z., Lawson, D.R. and Ashbough, L. (1994) Temporal and spatial variations of $PM_{2.5}$ and PM_{10} aerosol in Southern California. *Atmos. Environ.*, **28 (12)**, 2061–2081.

Clarke, A.G. and Karani, G.N. (1992) Characterisation of the carbonate content of atmospheric aerosols. *J. Atmos. Chem.*, **14**: 119–128.

Harrison, R.M. and Sturges, W.T. (1984) Physico chemical speciation and transformation reactions of particulate atmospheric nitrogen and sulphur compounds. *Atmos. Environ.*, **18**: 829.

Kerr, P.F. (1977) *Optical Mineralogy*. McGraw Hill, New York.

Lee, D.S., Dolland, G.J., Derwent, R.G. (1997) Observations on the partitioning of acids and bases between the gaseous and aerosol phases in the atmosphere. *Water, Air, Soil, Pollut.*. **113**: 175–202.

McCrone, W.C. (1978) *Identification of Asbestos by Polarized Light Microscopy*. National Bureau of Standards (US) Special Publication No. 506, Nov., US Food and Drug Administration, pp. 235–248.

McCrone, W.C. and McCrone, L. (1976) Light microscopy of micro-fibres – uses and limitations. In: *Symp. on Microscopy of Microfibers*. HEW Publications (FDA) No. 77-1033, US Food and Drug Administration, pp. 37–43.

Morgan, A.J., Davies, T.W. and Erasmus, D.A. (1975) Analysis of droplets from iso-atomic solutions as a means of calibrating a transmission electron analytical microscope. *J. Micros.*, **104**: 271–280.

Pooley, F.D. (1975) The identification of asbestos dust with an electron microscope microprobe analyser. *Ann. Occup. Hyg.*, **18**: 181–186.

Pooley, F.D. (1977) The use of an analytical electron microscope in the analysis of mineral dusts. *Phil. Trans. Roy. Soc. Lond. A.*, **286**: 625–638.

Pooley, F.D. (1981) Tissue mineral identification. In: *Occupational Lung Diseases: Research Approaches and Methods*. (eds. Weill, H. and Turner Warwick, M.) Marcel Dekker, New York, pp. 189–235.

Pooley, F.D. and Mille, M. (1999) Composition of Air Pollution Particles. In: *Air Pollution and Health*. (eds. Holgate, S.T., Samet, J.M., Koren, H.S. and Maynard, R.L.) Academic Press, pp. 619–634.

Review Group on Acid Rain (1997) *Acid deposition in the United Kingdom* 1992–94. AEA Technology, Culham, Oxford.

Rowse, J.B., Jepson, W.B., Bailey, A.T., Chimpson, N.A. and Soper, P.M. (1974) Composite elemental standards for quantitative electron microscope microprobe analysis. *J. Phys. E., Sci. Instrum.*, **7**: 512–514.

Watts, R., Cupitt, L. and Zweidinger, R. (1988) Mutagenicity of organics associated with $PM_{2.5}$ and PM_{10} Hi Vol particles from a wood smoke impacted residential area. In: *Int. Symp. EPA/APCA Measurement of Toxic and Related Air Pollut.* pp. 879–886.

Whei-May, Lee, G. and Lin-Yi-Tsay (1994) The partitioning model of polycyclic aromatic hydrocarbon between gaseous and particulate PM_{10} phases in urban atmosphere with high humidity. *Sci. Total Environ.*, **145 (1)**, 163–169.

4 The physicochemical characterization of urban airborne particulate matter

Kelly A. BéruBé, Timothy P. Jones, Ben J. Williamson and Roy J. Richards

4.1 Introduction

Airborne particulates are becoming an increasingly important environmental issue with immediate consequences for human health. They are mainly produced from natural sources, for example wind blown dust, smoke from forest fires, volcanic material, but under local urban conditions may be almost entirely derived from anthropogenic activities, especially from automobile exhaust emissions and smoke from power generation (Kouimtzis and Samara, 1995). Increasing industrialization and urbanization, with a resulting high density of road traffic has created a pattern of pollution distinctly different from that of surrounding rural areas. Urban air is characterized by ground-level pollution from carbon monoxide (CO), oxides of nitrogen (NO_x), benzene and particles. The ground-level sources of the pollutants means that they are not subject to effective dispersion processes and the effect on air quality is very substantial (Harrison, 1997). A comparison of the UK 1993 (nation-wide) emissions inventory of particulate matter with the 1990 Greater London Area inventory demonstrates the far greater proportion of emissions arising from road traffic within London as opposed to the country as a whole. In the Greater London Area approximately 87% (by weight) of primary PM_{10} (particulate matter with an aerodynamic diameter of $\leqslant$ 10 μm) emissions were derived from diesel exhaust emissions compared to 24% nation-wide (QUARG, 1996).

PM_{10} is currently regarded as the size fraction best representing those particles most likely to cause ill health (DoE, 1995). PM_{10} is not as long-

Particulate Matter, edited by R.L. Maynard and C.V. Howard

lived as $PM_{2.5}$, with a life-time of some 7–30 days, as the latter is less subject to efficient removal by gravitational settling or scavenging by rain (DoE, 1993). However, particles have to be < 2.5 μm (mean aerodynamic diameter) in order to penetrate into the gas exchange regions of the lungs. Numerous epidemiological studies have found a relationship between particulate air pollution and increased cardiorespiratory morbidity and mortality (Pope *et al.*, 1995), and hospital admissions for asthma and chronic obstructive pulmonary disease (Schwartz, 1994, Schwartz *et al.*, 1993). Electron microscope inspection of collection filters of PM_{10} reveal chain-like aggregates of spherical, ultrafine (UF) particulates (*Figure 4.1*). The number of particles in this size range is known to increase during air pollution episodes. Given their physical characteristics, it is not surprising that suspicion has centred on the UF fraction (0.001–0.1 μm mean aerodynamic diameter) of PM_{10} as being one of the primary agents in causing health effects during air pollution episodes (Richards, 1997).

The mechanistic explanation for the link between PM_{10} and adverse health effects in sensitive groups at low airborne mass concentrations (< 30 μg m^{-3} air) remains a mystery. The relative contributions of particle mass, size, chemical composition, or some combination of these factors, is unknown. Current research has focused on the chemical composition (transition metals, polyaromatic hydrocarbons [PAHs], nitro-PAHs, aero-allergens) and morphological properties (size, shape, aggregation properties) of particles as a means of understanding particle toxicity. Urban PM_{10} and diesel exhaust particulates (DEP) are themselves complex mixtures. In studying the toxicity of one of the components of these complex mixtures, toxicologists have used industrially manufactured carbon black (CB) particulates as a proxy for the refractory carbon of urban PM_{10} and DEP.

In this chapter, we discuss the physicochemical characteristics (e.g. physical dimensions, surface chemistry and solubility) of DEP, along with the 'control' particulates of CB. A range of microscopical and analytical techniques has been used including transmission electron microscopy (TEM), field emission scanning electron microscopy (FESEM), image analysis (IA), electron probe X-ray microanalysis (EPXMA), X-ray diffraction (XRD) and inductively coupled plasma mass and atomic emission spectrometry (ICP-MS; ICP-AES).

4.1.1 Trends in urban emissions

The domestic burning of coal in UK cities resulted in pollution episodes before and during the 1950s and 1960s that were characterized by high levels of sulphur dioxide and smoke 'particles' producing the infamous 'pea-souper' smogs. The Clean Air Act of 1956 reduced the source of these co-pollutants, but unfortunately they were to some extent replaced by an increasingly large component from motor vehicle emissions. Although in 1992 these were limited by government standards (DoE, 1996) effectively requiring the use of catalytic converters on all new petrol engine vehicles, the diminished emissions have been counteracted by an increased market share of diesel engine vehicles in the 1990s (QUARG, 1993). The perception that these vehicles are 'cleaner' and more 'environmental friendly' than petrol-based vehicles has changed. When compared with engines fitted with catalytic converters, which reduce CO, NO_2 and volatile organics, diesel engines emit less CO and hydrocarbons but more 'particles' and NO_2.

4.1.2 Diesel exhaust particulates

DEP are derived from the incomplete combustion of fuel and oil. Diesel exhaust emissions contain gases (CO_2, NO_2) and some PAH compounds in addition to the carbonaceous particles. The most troublesome diesel engine pollutant is the particle phase (*Figure 4.2a–d*), which consists of a variety of carbonaceous particles known as 'spherulites', 'chains' or 'clusters' (Armann and Siegla, 1982; BéruBé *et al.*, 1997; HEI, 1995; Vuk *et al.*, 1976), 'spherules' and 'flakes' (BéruBé *et al.*, 1997; Vuk *et al.*, 1976). The mass median aerodynamic diameter of DEP is approximately 0.2 μm, with 90% by mass less than 1 μm: the particles are thus readily respirable (HEI, 1995). The particles in diesel emissions have large surface areas

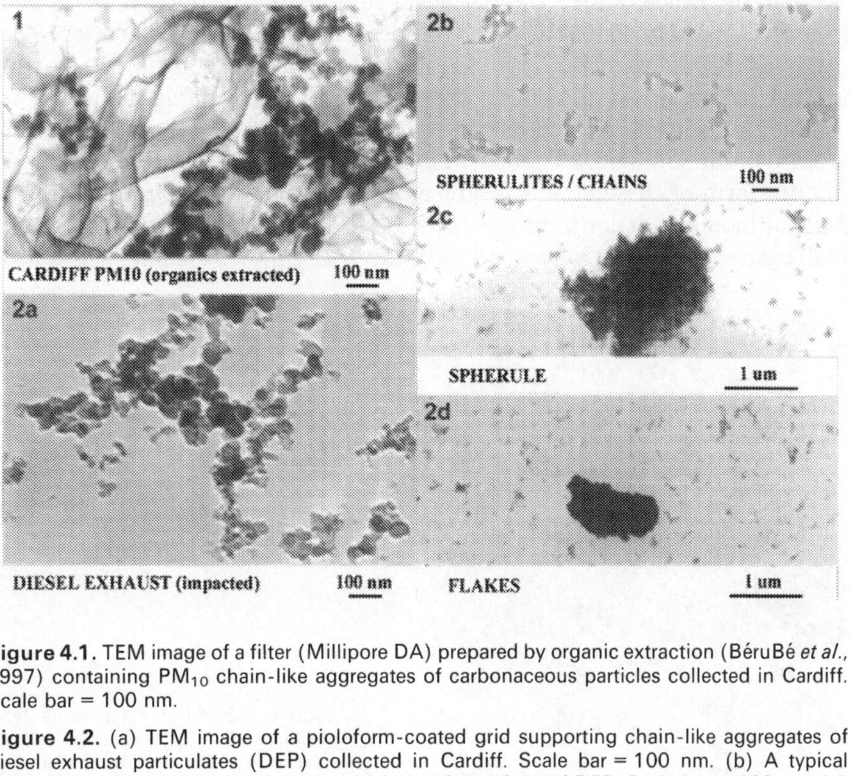

Figure 4.1. TEM image of a filter (Millipore DA) prepared by organic extraction (BéruBé *et al.*, 1997) containing PM_{10} chain-like aggregates of carbonaceous particles collected in Cardiff. Scale bar = 100 nm.

Figure 4.2. (a) TEM image of a pioloform-coated grid supporting chain-like aggregates of diesel exhaust particulates (DEP) collected in Cardiff. Scale bar = 100 nm. (b) A typical assemblage of 'spherulites', 'chains' and 'cluster' formations of DEP. Scale bar = 100 nm. (c) Agglomerated particles forming a 'spherule'. Scale bar = 1 μm. (d) Electron-dense 'flake-like body' found in DEP mixed with ambient urban air. Scale bar = 1 μm.

which allow the adsorption of organic compounds (typically 10–40% by mass) (HEI, 1995). Carcinogenic PAHs and nitroarenes attached to these respirable particles have been the focus of health concerns and the International Agency for Research on Cancer has classified DEP as a probable human carcinogen. However, identification of exactly which components of exhaust emissions, gases or particles, cause ill health is confounded by the fact that many of the compounds found in DEP are also emitted from other combustion sources.

4.1.3 Carbon black particulates

Carbonaceous material is a principal component of airborne particulate matter. It exists as either organic carbon or particulate elemental carbon, also known as carbon black. The International Union of Pure and Applied Chemistry (IUPAC) defines CB as an industrially manufactured carbon material composed of spheres and aggregates with a range of size between 10 and 1000 nm (LaHaye and Ehrburger-Dolle, 1994; *Figure 4.3*). It is generated during the incomplete combustion of various fuels, for example coal, gas, petrol and diesel oils, and in the UK in 1986, it was estimated that 17 ktons of CB were emitted into the atmosphere (Kouimtzis and Samara, 1995). Diesel combustion in the UK is thought to contribute between 80 and 95% of airborne CB levels (Colebeck, 1995).

CB is also generated from non-human sources such as forest fires ignited naturally by lightning. This material is classified either by its combustion sources and chemistry or by the analytical processes used for identification (Jones *et al.*, 1997). Thus, Goldberg (1985) defines CB as 'an impure form of the element produced by incomplete combustion of fossil fuels or biomass. It contains over 60% carbon with the accessory elements hydrogen, oxygen, nitrogen and sulphur. On the other hand, Kuhlbusch

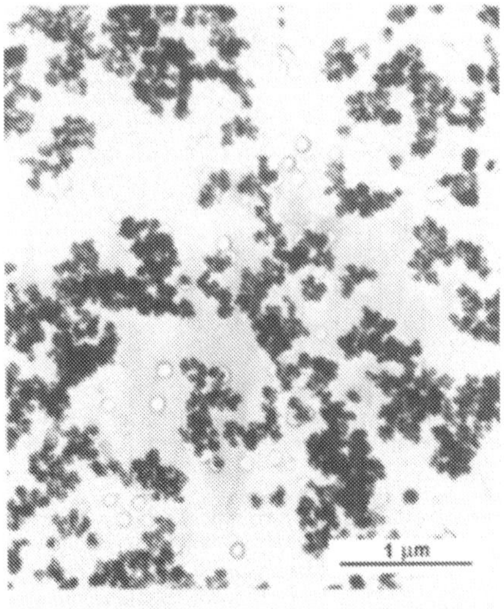

Figure 4.3. SEM image of a filter (Millipore DTTP) prepared by organic extraction containing chain-like aggregates of CB M120 particles. Scale bar = 1 μm.

and Crutzen (1995) consider CB to be 'a fire-derived highly aromatic to graphitic carbon fraction that is resistant at 340°C in pure oxygen and has a molar H/C-ratio of <0.2'.

4.2 Physical characterization of airborne particulates

The physical properties of particles determine the following biological responses (Fubini, 1996): (1) translocation into various biological compartments, and their fate *in vivo* (mediated by surface properties such as hydrophilicity and the presence of adsorbed substances), (2) activation of cells (related to surface chemistry and charge), (3) internalization within cellular compartments (dependent on shape and chemical composition), (4) potential free radical release, (5) duration of biopersistence/clearance. In order to evaluate the toxicity of airborne particulates, all of these characteristics and processes should be investigated.

4.2.1 Microscopy-based techniques

The microscopic nature of components of PM_{10} such as DEP and CB requires the use of either TEM or high resolution SEM to examine their physical characteristics such as size, shape and degree of particle aggregation. TEM can provide information about the fine structural features of particles by passing a beam of electrons through them, whereas SEM scans the surface of the particles with an electron beam and produces an image of the particle surface. These two types of electron microscopy therefore provide complimentary information about overall particle morphology. In addition to imaging by TEM or SEM, image analysis (IA) is used to obtain quantitative data from electron microscope images. IA is especially suitable for the quantitative assessment of particle morphological parameters which can be used to derive mass, volume and surface area data of particulate matter samples (BéruBé *et al.*, 1999).

4.2.2 Collection and preparation of DEP and CB

Collection of DEP. In our studies DEP are collected under dry, outdoor, ambient conditions from a diesel engine (Murphy *et al.*, 1998a). A high volume air sampler, operating at a flow rate of 30 l min^{-1} and attached to a total suspended particulate selective-inlet head is used to collect DEP as described in Murphy *et al.* (1998a). Collections are made on to pre-weighed polycarbonate filters (pore size 0.67 μm). These filters allow good capture rates of up to 20 mg h^{-1} (Pooley and Gibbs, 1996). PM_{10} collections are carried out under ambient outdoor conditions.

Suspended particles. The collection filter, containing a known weight of DEP, is added to a measured volume of pure grade water in a sterile, wide-bottomed, plastic container. All contact with metal objects, for example forceps or scrapers, is avoided and the container is placed for 15 min in an ultrasonic bath containing iced water. This procedure helps to 'wet' the hydrophobic DEP and removes the majority of particles from the filter. The filter is then removed and the suspended DEP sample diluted as required. Two microlitre samples of DEP suspensions are aliquoted on to pioloform-coated nickel TEM grids or on to SEM stubs and allowed to air-dry at room temperature in covered containers (BéruBé *et al.*, 1999).

Impacted particles. Impaction samples are defined as particulates that are deposited directly on to TEM grids or SEM stubs, without being wetted. This procedure allows the examination of particles in their natural state, since it is not necessary to disturb them post-collection. The pioloform-coated TEM grids and/or glass-cover slipped SEM stubs are mechanically secured by a purpose-built grid/stub holder attached to the exhaust outlet, and hence there is no need for the use of cements or glues: the risk of contamination is thus limited.

Preparation of CB. CB particulates referred to as Monarch 120, 880 and 900 and Regal 250 are obtained from the Cabot Corporation (USA), and are supplied as powdered preparations. Suspensions of these CBs are prepared as described above for DEP. They are also impacted by sedimentation on to SEM stubs coated with a thin layer of quick-curing Araldite epoxy resin. The resin is allowed to cure until hard but still sticky before particles are allowed to impact upon it. This prevents sinking of the particles beneath the surface of the resin.

Preparations for IA. Particles impacted on to TEM grids do not need any special preparation for IA. However, suspensions of particulate matter are processed through a vacuum filtration system using a glass frit filter support, which ensures uniform distribution of retained particles on the filter (Cyclopore Track Etch, Whatman; 0.2 μm pore size) surface. Filters are allowed to air-dry post-filtration in covered petri dishes, and later coated (1–2 nm) by evaporation of a carbon (graphitic) filament using a high vacuum coating unit (Peters, 1984) to provide a new substrate for particle adherence. The carbon-coated filters are then prepared for TEM as described in BéruBé *et al.* (1997). Briefly, areas the size of a TEM grid ($\sim$3 mm diameter) are cut away from the filter, placed on to a glass slide that is wetted with chloroform, then submerged in a petri dish containing chloroform. A blank TEM grid is then slid underneath the wetted filter and lifted out of the chloroform. The grid/filter is then placed in fresh chloroform for 5 min to allow the filter media to dissolve leaving the carbon coating to support the particles. The preparation is then thin enough for TEM.

A desktop image processing and analysis system (Leica, UK) is used to obtain quantitative measurements of particulate morphology. The primary measurements (in microns) are obtained from digitized images of TEM micrographs of particulate matter. Ten micrographs at a magnification of ×10 000 will show in the region of 1000 particles; a statistically acceptable population. Empirical measurements include the number of particles, Feret length, Feret breadth, area and perimeter. Derived parameters from these data include aspect ratio (the ratio of length to breadth) and roundness (a factor derived from the area and perimeter measurements). Calculations are made of equivalent spherical diameter (μm; ESD); a convention commonly used to define the size of particles with complex morphologies. From this, both particle surface area (μm^2) and volume (μm^3) are calculated. Mass (*M*; pg) can also be related to volume, where M=V2 and, for a given population of particles, the total surface area and mass can be calculated, as described by Bérubé *et al.* (1999).

Estimations were made of equivalent spherical diameter (ESD), and from this both particle surface area (*SA*, μm^2), and volume (*V*, μm^3) can be derived (*Equations 1* and *2*):

$$SA=4\pi[(\mathrm{ESD})/2]^2; \quad (1)$$

$$V=4/3\pi[(\mathrm{ESD})/2]^3. \quad (2)$$

If we assume a specific gravity (*SG*) for DEP of approximately 2 *SG*, that is intermediate between graphite (2.25) and turbostatic carbon black (1.86) [Cabot Corporation, USA], we can link mass (*M*, pg) and *V*:

$$M=V2. \quad (3)$$

And further derive a relationship between *M* and *SA* (derived from calculated ESD) and the number (*n*) of particles (*Equations 4* and *5*). Thus, for a given population (*n*) of particles, the total *SA* of *n* particles (SA_T) can be expressed as:

$$SA_T=\Sigma(4\pi[(\mathrm{ESD}_1)/2]^2):(4\pi[(\mathrm{ESD}_n)/2]^2). \quad (4)$$

For the same population (*n*) of particles, the total *M* of *n* particles (M_T):

$$M_T=[\Sigma([4/3]\pi[(\mathrm{ESD}_1)/2]^3):([4/3]\pi[(\mathrm{ESD}_n)/2]^3)]2. \quad (5)$$

Therefore, *SA* per $M=SA_T/M_T$. (6)

Three size ranges of airborne particles have been described. These are useful, not only for the purpose of classification, but also for a better understanding of the relative contributions of different-sized particulates in toxicological studies. The size ranges are referred to as the 'ultrafine' (UF) mode [0.001–0.1 μm], 'fine' (F) mode [0.1–2 μm] and 'coarse' (C) mode [2–10 μm] (QUARG, 1996). Particles larger than 10 μm in diameter are not considered as components of PM_{10}.

4.2.3 Transmission electron microscopy

Diesel exhaust particulates. Four different morphological types of DEP are consistently observed in samples collected from the exhaust outlet by impaction directly on to TEM grids. These being the individual particles or spherulites (mean diameter of 30 nm), small aggregates of spherulites in the form of chains or clusters, and large agglomerated spherules (1–2 μm in diameter) (*Figure 4.2a–c*). A fourth particle morphology that is less frequently observed has been referred to as flake-like bodies (0.5–2.5 μm in diameter) (*Figure 4.2d*). There are several possible sources for these flakes: (1) they may result from spherulite coalescence, (2) they may originate from the absorption and condensation of organic matter on to particles or (3) they may represent contaminants derived from ambient air trapped during sample collection. The aggregation properties of DEP collected by impaction are comparable morphologically to the soot fraction of PM_{10} from ambient, urban air. The morphology of DEP suspensions, that are removed from collection filters via sonication, is generally the same as that of the impacted samples. One important difference lies in a decrease in the number of particles in the ultrafine population found in the suspensions. This could be a function of the different preparation protocols.

Carbon black particulates. CB particulates prepared as suspensions exhibit three morphological types similar to DEP, for example spherulites, chains/clusters and spherules. They exhibited different aggregation properties which can be described as being either 'low structured' blacks, that is containing few spherulites per aggregate (e.g. M120), or 'high structured' blacks, that is containing many spherulites per aggregate (e.g. R250). CB Monarch 120 particles (spherulite mean diameter of 50 nm) form simple aggregates similar to DEP, characterized by chains and a noticeable lack of spherules within the particle population (*Figure 4.3*). In contrast, the CB Regal R250 particles (spherulite mean diameter of 40 nm), exhibit a high degree of aggregation (plaque-like), with a tendency towards larger particle (spherules) formation (*Figure 4.4*).

4.2.4 Scanning electron microscopy

Diesel exhaust particulates. The SEM examination of DEP supports the results from TEM in that the particle morphologies can be broadly classified as spherulites, chains/clusters, spherules and 'flake-like' bodies. However, when compared with DEP from particle suspensions, impacted DEP samples collected directly on to SEM stubs exhibited a different particle organization (*Figure 4.5a* and *b*). Observations of wetted/ sonicated DEP (*Figure 4.5a*) exhibit 'smooth', compact-type aggregates with apparently smaller surface areas and fewer discrete ultrafine particles, whereas impacted or 'dry' DEP (*Figure 4.5b*) reveal that particles consist of 'flocculated', cluster-type aggregates with apparently

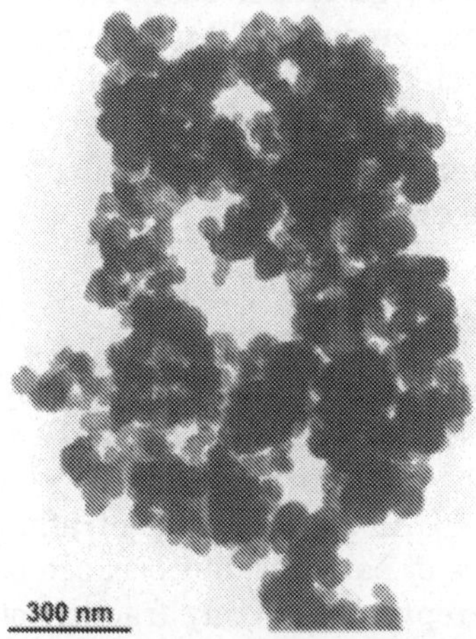

Figure 4.4. TEM image of a pioloform-coated grid supporting 'plaques' of CB R250 aggregated particles. Scale bar = 300 nm.

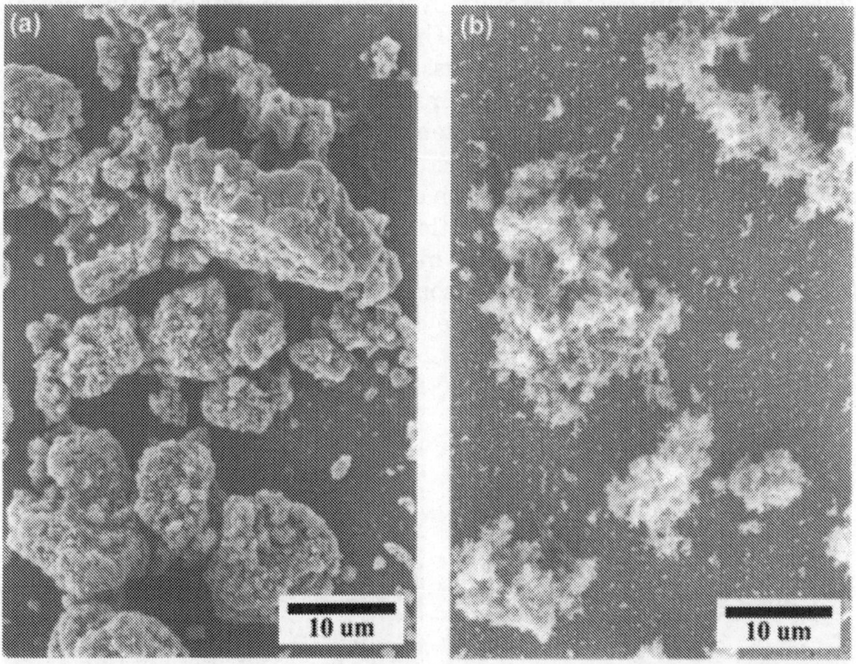

Figure 4.5. Field emission scanning electron microscope images comparing sonicated versus impacted diesel exhaust particulates (DEP). (a) Sonicated DEP exhibiting a compact arrangement of spherulites. Note reduced number of ultrafine/fine particles in the background. (b) Impacted DEP revealing flocculated arrangement of spherulites. Note significant number of ultrafine/fine particles in the background. Scale bars = 10 μm.

large surface areas and a large population of ultrafine particles. The combination of wetting and sonication appears to compress the original open-spaced packing arrangement of the particles to yield a more 'rounded' particle appearance.

Carbon black particulates. The same particle organization observed with DEP for sonicated or 'wet' preparations is encountered with CB particulates, that is, they form round and compact aggregates (*Figure 4.6a*). Similarly, CB collected directly on to SEM stubs present the same arrangement as their impacted DEP counterparts. These particulates presented flocculated aggregates of potentially large surface areas (*Figure 4.6b*).

4.2.5 Quantitative image analysis

Diesel exhaust particulates. The distributions of sonicated DEP, by number, mass and surface area relative to ESD, derived from IA data are shown in *Figure 4.7a–c*. Importantly, all the particles are in the 'respirable' range (i.e. <2.5 μm in diameter; QUARG, 1996). The ultrafine mode (0.001–0.1 μm) contains large numbers of particles, but because of their relatively small size this component represents a tiny fraction of the total mass. In contrast, flakes and spherules in the coarse mode (2.0–10 μm) contribute substantially to the mass, although the actual number of these particles is small. The two different ways of interpreting the data, by either number or mass, confer greater significance to the contribution of the smaller or larger particles, respectively. The number distribution emphasizes the importance of the ultrafine and fine (0.1–2.0 μm) particles, whereas the mass distribution emphasizes the importance of the coarse particles. A comparison of the distributions of DEP prepared by wetting/sonication versus impaction is shown in *Figure 4.8*. Sonication causes the number of particles in the ultrafine size range to fall from 60.87% (impacted) to 10.1% (sonicated) and the number of particles in the fine size range to increase from 39.13% to 89.5% and in the

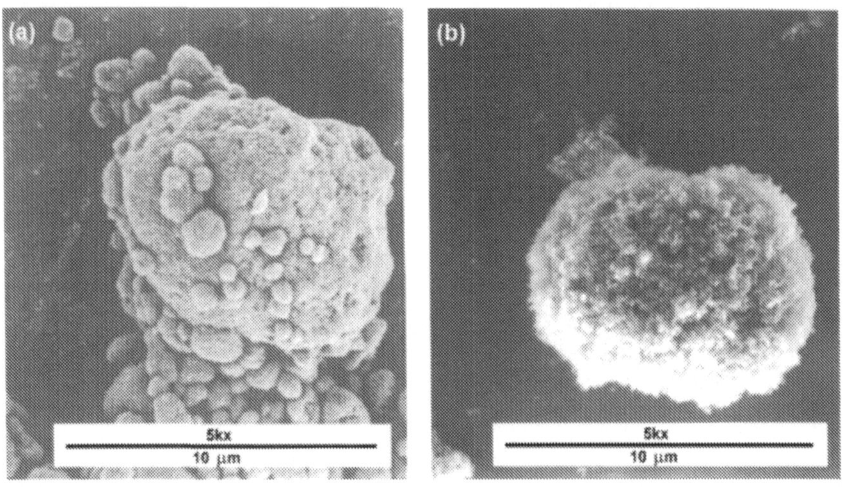

Figure 4.6. Field emission scanning electron microscope images comparing sonicated versus impacted carbon black (CB) particulates. (a) Sonicated CB M120 exhibiting a compact arrangement of spherulites. (b) Impacted CB M120 revealing flocculated arrangement of spherulites. Scale bars = 10 μm.

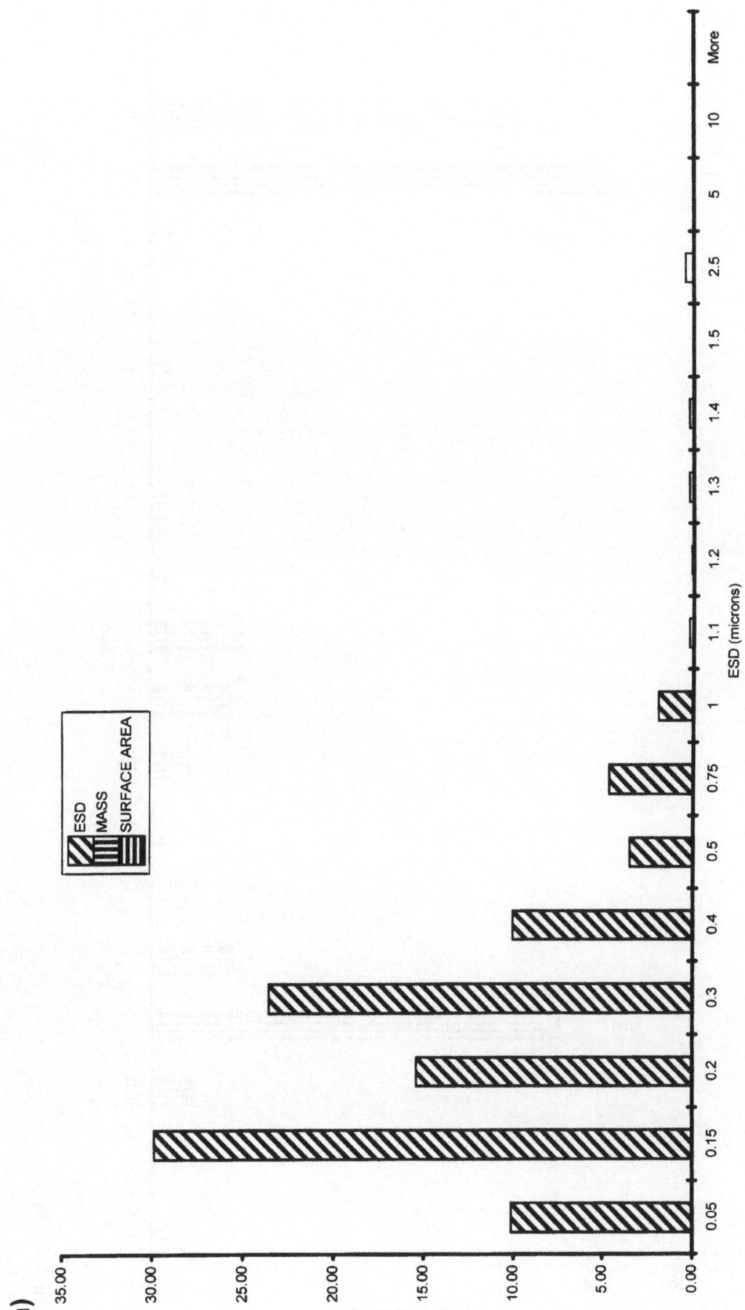
(a)
Percentage
35.00
30.00
25.00
20.00
15.00
10.00
5.00
0.00
ESD
MASS
SURFACE AREA
0.05
0.15
0.2
0.3
0.4
0.5
0.75
1
1.1
1.2
1.3
1.4
1.5
2.5
5
10
More
ESD (microns)

Figure 4.7. (a)

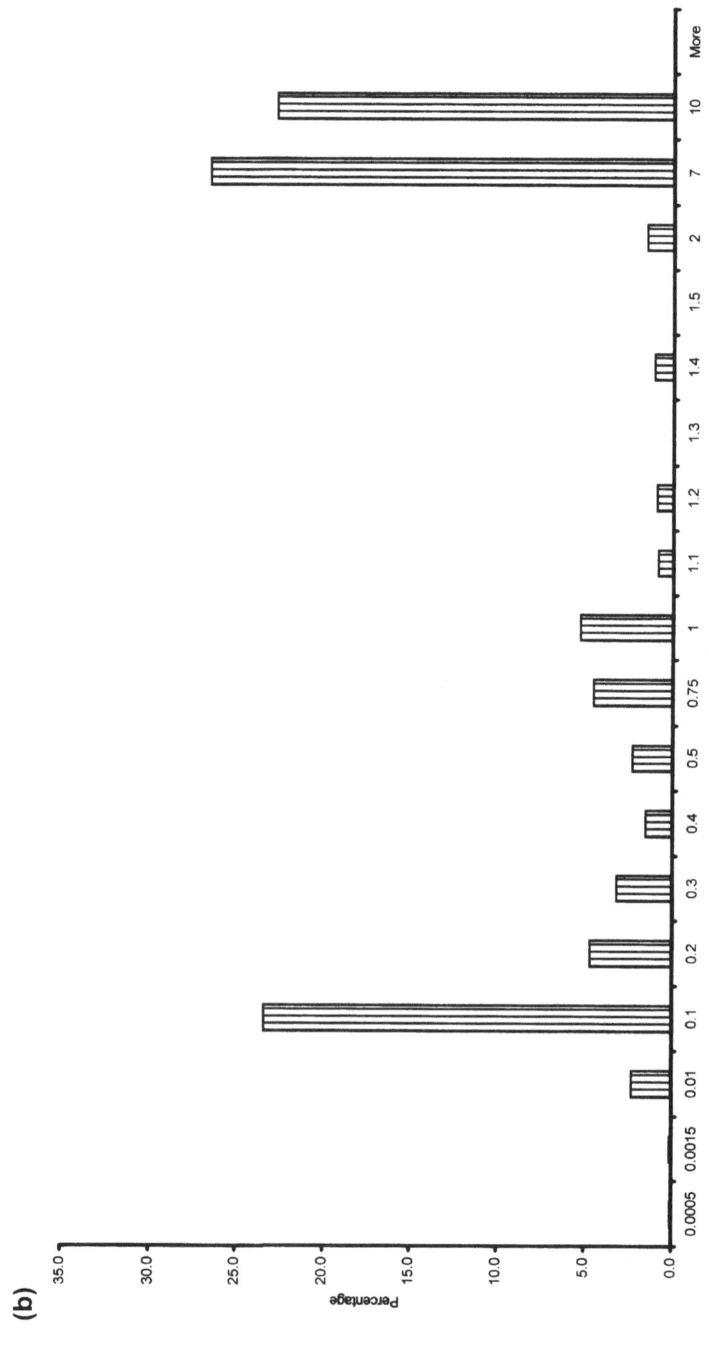
(b)
Percentage
35.0
30.0
25.0
20.0
15.0
10.0
5.0
0.0
Picograms
0.0005
0.0015
0.01
0.1
0.2
0.3
0.4
0.5
0.75
1
1.1
1.2
1.3
1.4
1.5
2
7
10
More

Figure 4.7. (b)

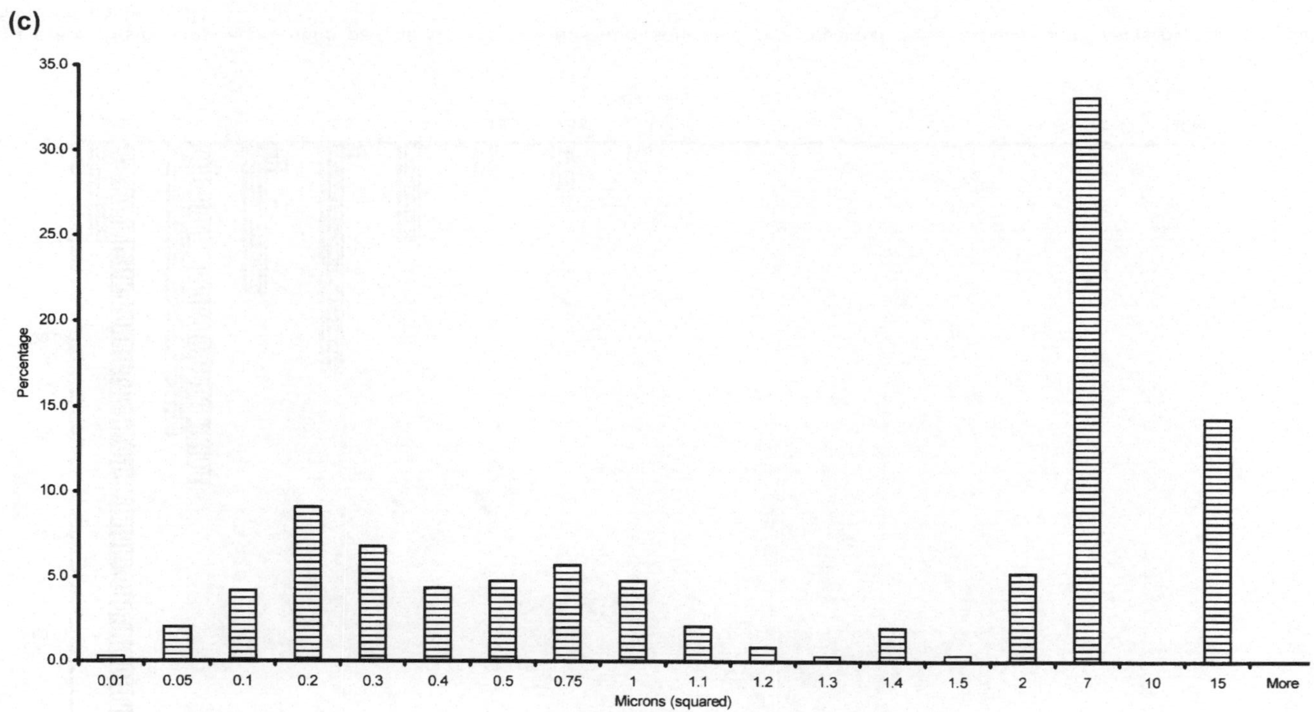

Figure 4.7. Histograms depicting particle size distributions from wetted/sonicated preparations of diesel exhaust particulates. (a) Number; (b) mass; (c) surface area. ESD, equivalent spherical diameter.

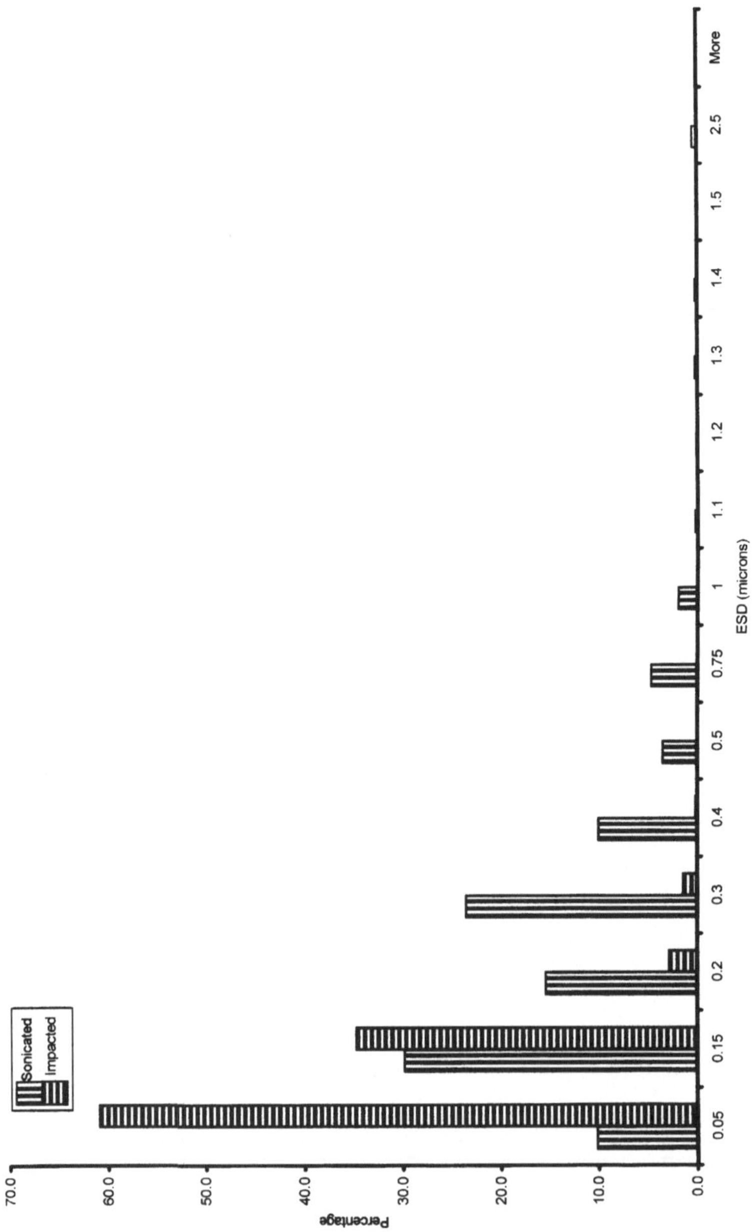

Figure 4.8. Histograms representing particle size distributions from sonicated and impacted diesel exhaust particulates by number relative to equivalent spherical diameter (ESD).

coarse size range from 0% to 0.4%. Nevertheless, both sample types are respirable.

When comparing the effects of different types of particles, the question arises as to how to express exposure or dose in toxicological studies. Mass is often used, but alternatives include the number of particles, the number of sized particles and the unit surface area. The appropriate parameter depends on the biological process under investigation. For example, exposed surface area would be the most appropriate parameter if small molecules are thought to be acting as mediators of biochemical reactions taking place (Fubini, 1996). The different DEP distributions produced from ESD data and the derived values, surface area and mass, demonstrate that great caution must be exercised when choosing the most appropriate way of presenting data. The advantages of using the number of particles of given ESD is that it is an empirical value, representing a commonly used parameter which conforms to particle legislation (aerodynamic diameter is approximately equivalent to ESD), and when using IA, very large data sets with low standard deviations are quickly produced. The drawbacks are that number places emphasis on the smaller particles, the complexity of particle morphology is disregarded, and toxicological doses are generally administered by mass. The benefit of calculating reactive surface area is that the interface between the particles and the lung is considered; this may provide a better estimate of toxicity than mass alone. However, DEP particles are so morphologically complex that the surface area calculations based on ESD must significantly underestimate the real surface area (see BéruBé *et al.*, 1999). These same principals apply for other morphologically complex particles such as CB and the soot fraction of PM_{10}.

Carbon black particulates. Histograms comparing the distributions of sonicated preparations of a low and a high structured CB, by number, mass and surface area relative to ESD, are shown in *Figure 4.9a* and *b*. The majority of particles from both samples are in the fine category (0.1–2 μm), as defined by ESD. The low structured CB M120, that is low aggregation, comprises predominately fine particles with a limited number of coarse particles (*Figure 4.9a*). The same situation as found in DEP is seen here in CB, that is, the ultrafine fraction (0.001–0.1 μm), because of its relatively small size, does not significantly contribute to the total mass and moderately so towards the surface area. The degree of particle aggregation will also affect the resultant particle size distribution. This effect is especially evident in the high structured CB R250. Although R250 has nearly the same spherulite diameter as M120, its propensity to form complex aggregates skews its distribution towards the coarse fraction (2.0–10 μm) (*Figure 4.9b*). This phenomenon is also seen with the other high structured CBs M880 and M900.

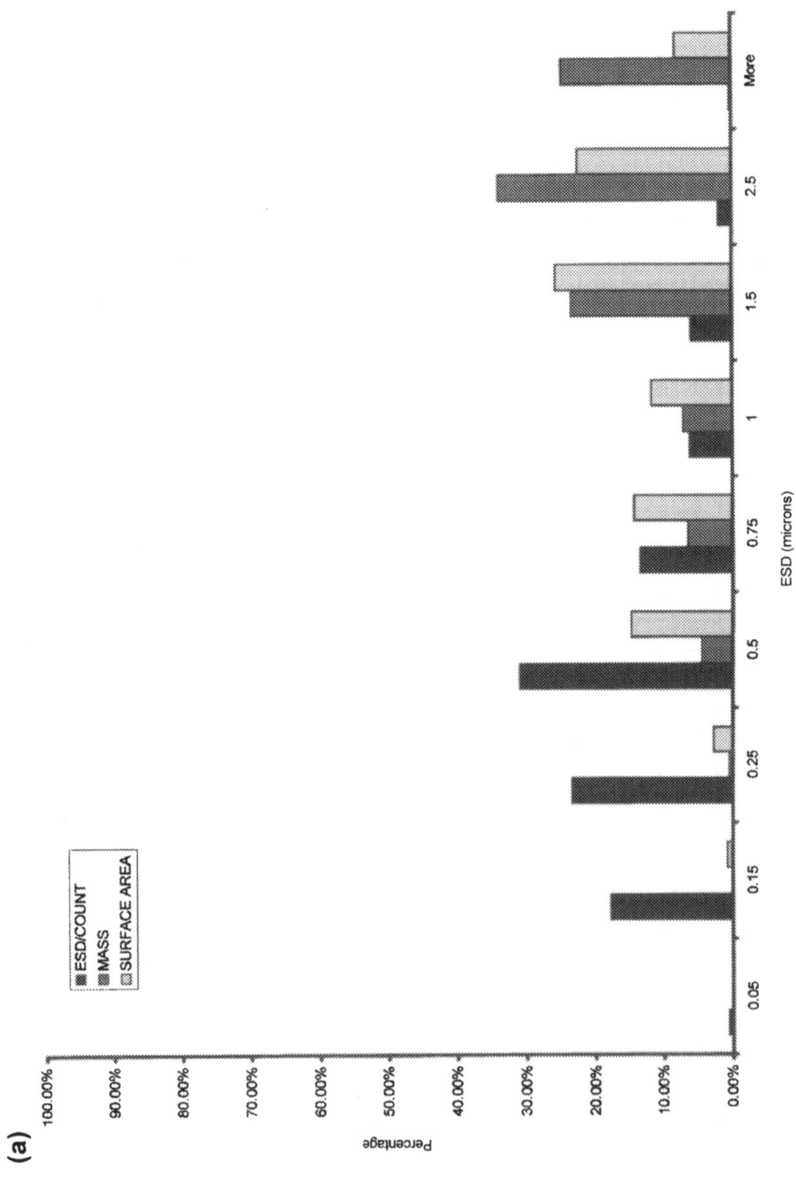
(a)
Percentage
100.00%
90.00%
80.00%
70.00%
60.00%
50.00%
40.00%
30.00%
20.00%
10.00%
0.00%
ESD/COUNT
MASS
SURFACE AREA
0.05
0.15
0.25
0.5
0.75
1
1.5
2.5
More
ESD (microns)

Figure 4.9. (a)

(b)

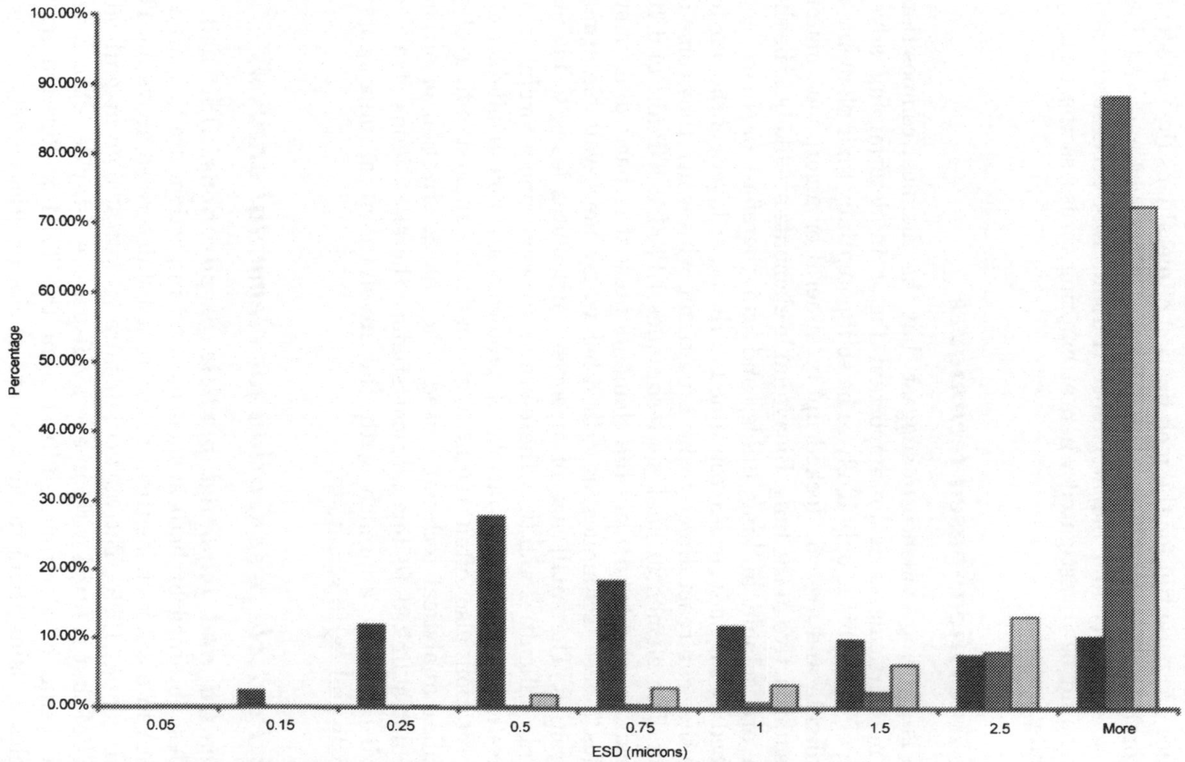

Figure 4.9. Histograms demonstrating particle size distributions produced by wetting/sonicating carbon black (CB) particulates, as assessed by number, mass and surface area relative to equivalent spherical diameter (ESD). (a) CB M120 particulates; (b) CB R250 particulates.

4.3 Chemical characterization of airborne particulates

The chemical properties of particles determine biological responses by: (1) adsorption of exogenous material (the particle may act as a carrier), (2) adsorption of endogenous material (the particle may be opsonized or inactivated), (3) potential for free radical generation at solid–liquid interface, (4) potential for the mobilization of metal ions (by endogenous chelators) and (5) by a propensity for selective leaching, and hence variable solubility and biopersistence (Fubini, 1996). These characteristics should be investigated when attempting to assess the toxicity of particulate matter.

4.3.1 Analytical-based techniques

There is growing evidence to suggest that the chemical composition of airborne particulates is an important factor in determining adverse health effects. This is only likely to be so if potentially toxic elements are bioavailable and are not 'locked up' in mineral or amorphous material that is unreactive in the lung. Bioavailable elements are likely to be those which are soluble in lung fluids, and can therefore be more easily absorbed by the body, or are contained in the outer layers of the particles where they can react directly with lung tissue with which they come into contact. The elemental analysis techniques EPXMA, ICP-MS and ICP-AES can be used to identify the elements present in particulate matter samples. These techniques, described below, are now available in many universities. The small size of airborne particulates leads to immense handling difficulties and they therefore require extremely careful sample preparation, for which a variety of protocols are now available. Data interpretation also often requires an in-depth understanding of the particular analytical technique used. The characterization of airborne particulates linked to toxicological studies therefore lends itself to a multidisciplinary approach with the involvement of microscopists, chemists and earth scientists.

4.3.2 Particle preparation for elemental analysis

Suspended and impacted particle preparations. DEP and CB samples, obtained by both suspension and impaction, are prepared for EPXMA in a manner identical to the protocol described above for TEM (BéruBé *et al.*, 1999). The particulates are viewed at high magnification in a TEM that is fitted with an energy dispersive X-ray detector. Briefly, a fine beam of electrons emitted from the TEM is directed at the particles causing them to emit X-rays of characteristic energies for all elements present.

Particle-free suspensions. Suspensions of particulate matter, prepared using techniques described previously, are used for ICP-MS and ICP-AES. These preparations are sonicated to ensure homogeneous 'washing' of the particles, ultracentrifuged (50 000 ***g*** for 30 min) to separate particles and the supernatant decanted into sterile tubes according to BéruBé *et al.* (1999). The liquid fraction is examined by TEM to ensure that it is particle-free prior to analysis. The particle-free supernatant is then analysed using an inductively coupled plasma mass spectrometer to determine the concentrations of most of the elements in the periodic table or atomic emission spectrometer for transition, alkali and alkaline earth metals not detected by ICP-MS, following standard techniques (Jarvis *et al.*, 1992).

4.3.3 Electron probe X-ray microanalysis

Diesel exhaust particulates. Electron probe X-ray microanalysis of DEP reveals that their chemical compositions can differ between morphological types, with duration of collection time and with preparation conditions, that is impaction or sonication. Static probe analysis of the basic morphological types of particles from impacted samples of DEP reveals that small chains and clusters of spherulites in the submicron range have the simplest chemistry (*Table 4.1*). Spherules initially exhibit similar elemental compositions to those of the chains and clusters, but with lengthy collection times, phosphorus and calcium are detected. The flake-like bodies always contain a wider range of elements, characterized by a heterogeneous mixture of transition metals. Several of the inorganic elements present in DEP are attributed to additives in the engine lubricating oil, as well as being part of the diesel fuel components. Zinc, phosphorus and calcium are common components of oil additives and sulphur, along with traces of calcium, iron, silicon and chromium can be found in the fuel.

Table 4.1. Electron probe X-ray microanalysis comparison of the different classes of diesel exhaust particulate prepared by impaction and sonication versus time

IMPACTED	Chains	Spherules	Flake-like bodies
Time			
30 sec	C, O, Si, S	C, O, Si, S	C, O, Na, Mg, Al, Si, S, K, Ca
1 min	C, O, Si, S	C, O, Si, S	C, O, Na, Mg, Al, Si, P, S, K, Ca, Fe, Zn
3 min	C, O, Si, S	C, O, Si, S	C, O, Na, S, P, Cl, K, Ca, Zn
5 min	C, O, Si, S	C, O, Si, S	C, O, Na, Mg, Al, Si, P, S, K, Ca, Fe, Zn
7 min	C, O, Si, S	C, O, Si, S, P, Ca	C, O, Na, Mg, Si, P, S, Ca, Zn
9 min	C, O, Si, S	C, O, Si, S, P, Ca	C, O, Na, Mg, Al, Si, P, S, K, Ca, Ti, Fe, Zn
11 min	C, O, Si, S	C, O, Si, S, P, Ca	C, O, Na, Mg, Si, P, S, Ca, Zn, Fe
13 min	C, O, Si, S	C, O, Si, S, P, Ca	C, O, Na, Mg, Al, Si, P, S, K, Ca, Fe, Zn
15 min	C, O, Si, S	C, O, Si, S, P, Ca	C, O, Na, Mg, Si, P, S, Cl, Ca, Ti, Fe, Zn
17 min	C, O, Si, S	C, O, Si, S, P, Ca	C, O, Na, Mg, Al, Si, P, S, Ca, Ti, Fe, Zn

SONICATED	Chains	Spherules	Flake-like bodies
Time			
15 min	C, O, Na, Mg, Al, Si, P, S, Ca, Fe, Zn	C, O, Na, Mg, Al, Si, P, S, Cl, K, Ca, Fe, Zn	C, O, Na, Mg, Al, Si, P, S, Cl, Ca, Fe, Zn

The process of wetting/sonication induces the translocation of soluble elements and their subsequent re-distribution on to other particles (*Table 4.1*). Comparisons by EPXMA of the two preparation procedures, that is wet and dry, demonstrate that in the wetted samples of DEP, small chains/clusters, spherules and flakes contain the same spectrum of elements. In the dry samples of DEP, discrete chemical signatures can be assigned to each type of particle.

Carbon black particulates. The low structured CB particles M120 possess a very simple composition when compared with their high structured counterparts (*Table 4.2*). CB M120 has a very similar elemental profile to the spherulites, chains and clusters that comprise DEP; this finding supports their use as a surrogate control particle for DEP. The high structured CBs M900 and M880 exhibit similar but more complex chemical signatures and, compositionally, these particles are more akin to the spherules found in DEP samples. Regal 250 is the only CB to contain a transition metal, iron, in addition to the common elements carbon, oxygen, silicon and sulphur. These differing chemical compositions between the four types of CB particulates, along with particle size, have been shown to result in differential toxic effects *in vitro* (BéruBé *et al.*, 1998; Murphy *et al.*, 1998b).

X-ray mapping. Qualitative X-ray mapping techniques, also referred to as 'speed-mapping' (BéruBé *et al.*, 1999; Morgan *et al.*, 1994), are employed to determine the relative spatial distributions of elements in the structurally heterogeneous sampling fields of particulate matter (*Figure 4.10*). Speed-mapping permits the correlation of structural information with chemical content and has demonstrated chemical heterogeneity within particulate matter samples. The utilization of this technique has shown that for DEP, the chemical compositions of the different classes of particles (e.g. chains, spherules, flakes), are not

Table 4.2. Electron probe X-ray microanalysis comparison of the low and high structured carbon black particulates

Structure	Carbon Black	Elements
Low	M120	C, O, Si, Cl
High	M880	C, O, Na, Si, S, Cl, K, Ca
	M900	C, O, Na, Si, S, Cl, K
	R250	C, O, Si, Cl, Fe

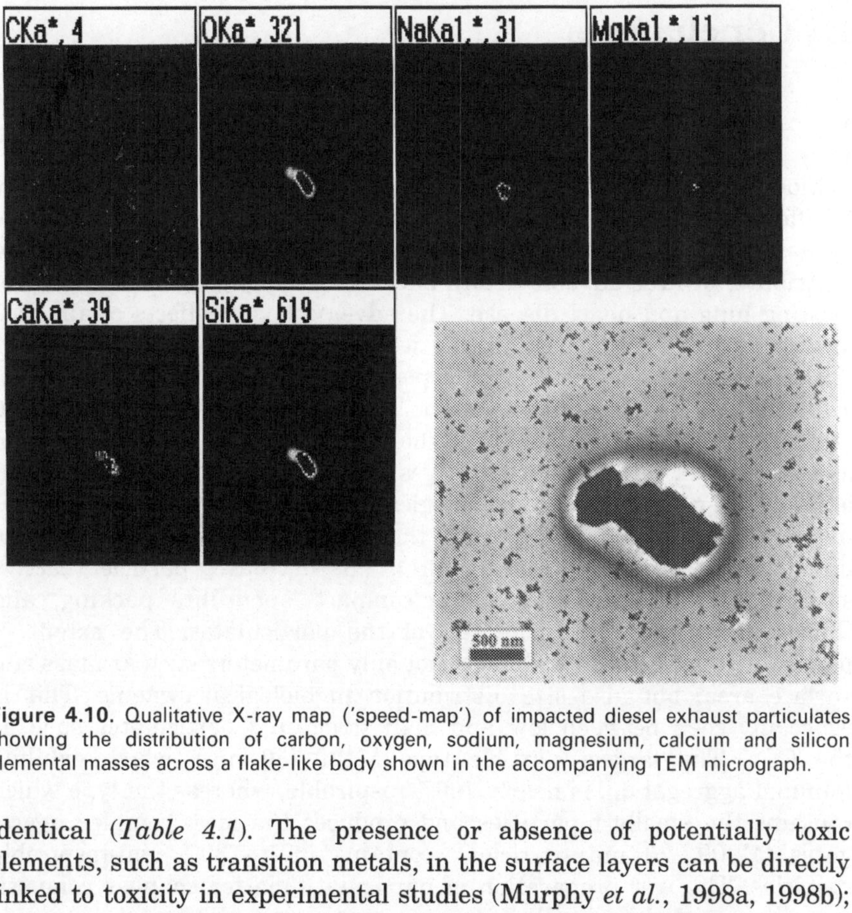

Figure 4.10. Qualitative X-ray map ('speed-map') of impacted diesel exhaust particulates showing the distribution of carbon, oxygen, sodium, magnesium, calcium and silicon elemental masses across a flake-like body shown in the accompanying TEM micrograph.

identical (*Table 4.1*). The presence or absence of potentially toxic elements, such as transition metals, in the surface layers can be directly linked to toxicity in experimental studies (Murphy *et al.*, 1998a, 1998b); moreover, particle 'chemical' bioavailability can be estimated.

4.3.4 Inductively coupled plasma mass and atomic emission spectrometry

Diesel exhaust and carbon black particulates. The chemical composition of particulate matter in suspension is altered by the 'wetting' process. The elements found above detection limits in the particle-free supernatants of DEP and which have therefore been dissociated from the particles, are magnesium, phosphorus, calcium, chromium, manganese, zinc, strontium, molybdenum, barium, sodium, iron, sulphur and silicon. The elements in concentrations found above detection limits in the four CB particle-free supernatants are very similar to those of DEP. The exception is the absence of phosphorus, zinc and molybdenum in any of the CB supernatants and the absence of vanadium in any DEP supernatants. All the elements present may either remain in solution or be sorbed on to other particles, which is suggested by EPXMA static probe analysis of wet particles versus dry particles.

4.4 Conclusions

Microscopic airborne particulates are ubiquitous in our environment. They may be produced by human activities, intentionally and unintentionally, or by natural processes. Many such particles have the ability to affect our health and quality of life. Particles that are capable of bypassing the natural defence processes of the lung are considered to contribute towards adverse health effects especially in persons with pre-existing lung and heart disease. The adverse health effects of airborne particulates depend upon their physicochemical properties.

These studies suggest that the post-collection preparation of urban airborne particulate matter such as DEP and CB using standardized methods for toxicological assays, that is delivery in a liquid medium and sonication, can introduce changes in the physicochemical characteristics of the particulates. Morphologically, these preparation techniques increase aggregation and reduce the number of particulates in the ultrafine/fine fractions. Consequently, this decreases particle reactive surface area by producing more 'compact' spherulite packing, and thus reduces the toxic potential of the particulates. The extent of particulate aggregation will affect not only parameters such as mass and surface area, but also size distribution in biological systems. This is especially seen between low and high structured particulates such as the CBs. The largest particle type M120 (50 nm), which exhibits minimal aggregation, is nearly 100% respirable, whereas the type which contains the smallest particles and produces the most complex aggregates (M900, 20 nm particles), contains up to 30% non-respirable particles. The wet preparation of particulate matter changes compositional chemistry as well. It promotes the dissolution of inorganic elements producing a wider spectrum of elements potentially available for chemical reaction; in dry samples the non-solubilized elements may not be immediately accessible for reaction. Nevertheless, these changes may well take place in humid climates or upon respiratory tract deposition (Colebeck *et al.*, 1990; Huang *et al.*, 1994). Therefore, when investigating the potential toxicity of mostly insoluble particles such as DEP and CB, beside mass, other parameters of toxicological significance such as particle number, surface area, shape, size, surface reactivity and chemistry are likely to be of importance for characterizing and/or interpreting any biological responses.

Acknowledgements

The work carried out for this publication was supported by the Medical Research Council, UK and a Cardiff University (Wales) Research

Fellowship for Dr Kelly A. BéruBé. The authors also wish to thank Chris Jones (Head of EM Unit, Natural History Museum, London) for assistance with FESEM.

References

Armann, C.A. and Siegla, D.C. (1982) Diesel particles—What are they and why. *Aerosol Sci. Technol.* **1**: 73–101.

BéruBé, K.A., Jones, T.P. and Williamson, B.J. (1997) Electron microscopy of urban airborne particulate matter. *Microsc. Anal.* **61**: 11–13.

BéruBé, K.A., Murphy, S.A. and Richards, R.J. (1998) Effects of carbon black physicochemistry on primary isolates of lung epithelial cells. *Eur. Resp. J.* **12** (Suppl. 28): A335.

BéruBé K.A., Jones T.P., Williamson B.J., Winters C., Morgan A.J. and Richards R.J. (1999) Physicochemical characterization of diesel exhaust particles: factors for assessing biological activity. *Atmos. Environ.* **33** (10): 1599–1614.

Colebeck, I. (1995) Particle emission from outdoor and indoor sources. In: *Airborne Particulate Matter* (eds T. Kouimtzis and C. Samara). Springer-Verlag, Berlin, pp. 1–35.

Colebeck, I., Appleby, L., Hardman, E.J. and Harrison, R.M. (1990) The optical properties and morphology of cloud-processed carbon smoke. *J. Aerosol Sci.* **21** (4): 527–538.

Department of the Environment (1993) *Urban Air Quality in the United Kingdom: First Report of the Quality of Urban Air Review Group.* HMSO, Bradford.

Department of the Environment (1995) *Expert Panel on Air Quality Standards: Particles.* HMSO, London.

Department of Environment (1996) *Digest of Environmental Statistics.* HMSO, London.

Fubini, B. (1996) Use of physico-chemical and cell free assays to evaluate the potential carcinogenicity of fibres. In: *Mechanisms of Fibre Carcinogenicity* (eds A.B. Kane, P. Boffetta, R. Saracci and J.D. Wilbourn). IARC Scientific Publications, Lyon, pp. 35–55.

Harrison, R.M. (1997) Urban air pollution in the United Kingdom. In: *Air Pollution in United Kingdom* (eds. G. Davison and C.N. Hewitt). The Royal Society of Chemistry, Lancaster, pp. 22–39.

Haung, P.F., Turpin, B.J., Pipho, M.J., Kittelson, D.B. and McMurray, P.H. (1994) Effects of water condensation and evaporation on diesel-chain agglomerate morphology. *J. Aerosol Sci.* **25** (3): 447.

Health Effects Institute (1995) *Diesel Exhaust: A Critical Analysis of Emission, Exposure and Health Effects. A Special Report of the Institute's Diesel Working Group.* HEI, Massachusetts.

Jarvis, K.E., Gray, A.L. and Houk, R.S. (1992) *Handbook of Inductively Coupled Plasma Mass Spectrometry.* Blackie, Glasgow.

Jones, T.P., Chaloner, W.G. and Kuhlbusch, T.A.J. (1997) Proposed bio-geological and chemical based terminology for fire-altered plant matter. In: *Sediment Records of Biomass Burning and Global Change*, Vol. 51 (eds J.S. Clark, H. Cachier, J.G. Goldammer and B. Stocks). NATO ASI Series 1: Global Environmental Change. Springer, Berlin, pp. 9–22.

Goldberg, E.D. (1985) *Black Carbon in the Environment.* Wiley, New York.

Kouimtzis, T. and Samara, C. (1995) *Airborne Particulate Matter: The Handbook of Environmental Chemistry*, Vol. 4, Part D (ed. O. Hutzinger). Springer-Verlag, Berlin.

Kuhlbusch, T.A.J. and Crutzen, P.J. (1995) Towards a global estimate of black carbon in residues of vegetation fires representing a sink of atmospheric CO_2, and a source of O_2. *Global Biogeochem. Cycles* **9** (4): 491–506.

LaHaye, J. and Ehrburger-Dolle, F. (1994) Mechanisms of carbon black formation: correlation with the morphology of aggregates. *Carbon* **32** (7): 1319–1324.

Morgan, A.J., Winters, C. and Yarwood, A. (1994) Speed-mapping of arsenic distributions in the tissue of earthworms inhabiting arsenious soil. *Cell Biol. Inter.* **18** (9): 911–914.

Murphy, S.A., BéruBé, K.A., Pooley, F.D. and Richards, R.J. (1998a) The response of lung epithelium to well characterized fine particles. *Life Sci.* **62** (19): 1789–1799.

Murphy, S.A., BéruBé, K.A. and Richards, R.J. (1998b) Effects of four carbon black samples and diesel exhaust particulates on isolated type 2 and clara cells. *Proc. Royal Microsc. Soc.* **33** (2): A113.

Peters, K.R. (1984) Precise and reproducible deposition of ultra thin carbon films by flash evaporation of carbon yarn in high vacuum. *J. Microsc.* **133:** 17.

Pooley, F.D. and Gibbs, A.R. (1996) *Characterisation of Fine Particulate Matter (PM_{10}) in Airborne Samples and Tissues.* Department of Health, London.

Pope, C.A., Bates, D.V. and Raizenne, M.E. (1995) Health effects of particulate air pollution: time for re-assessment? *Environ. Health Perspect.* **103**: 472–480.

Quality of Urban Air Review Group (1993) *Diesel Vehicle Emissions and Urban Air Quality.* Department of Environment, London.

Quality of Urban Air Review Group (1996) *Airborne Particulate Matter in the United Kingdom. Third Report.* Department of Environment, London.

Richards, R.J. (1997) Small particles, big problems? *Biologist* **44** (1): 249–251.

Schwartz, J. (1994) What are people dying of on high air pollution days? *Environ. Res.* **64**: 26-35.

Schwartz J., Slater D. and Larson T.V. (1993) Particulate air pollution and hospital emergency visits for asthma in Seattle. *Am. Rev. Resp. Dis.* **147**: 826–831.

Vuk, C.T., Jones, M.A. and Johnson, J.H. (1976) The measurement and analysis of the physical character of diesel particulate emissions. *Soc. Auto. Engin. Trans.* **85**: 556–589.

5 The structural and physical chemistry of nanoparticles

D.A. Jefferson and E.E.M. Tilley

5.1 Introduction

Nanoparticles are rather loosely defined as agglomerates of matter with dimensions less than 10^{-8} μm. Although they constitute a proportion of almost all types of particulate material, their importance has hitherto only been appreciated in the field of heterogeneous catalysis, where their high surface area has a very great influence on reactivity. Generally, they seem to be regarded as being merely extremely small crystals of inorganic materials, but at present we know little about their detailed structure, or their chemical and physical properties. In a way this is not surprising, as methods of preparing nanoparticles are still relatively crude, producing assemblages which inevitably show a very broad distribution of particle sizes. Any attempt at probing their properties using spectroscopic techniques such as infra-red absorption or magnetic resonance methods is therefore bound to be extremely difficult, as only a very approximate average of the properties of individual particles is obtained, with those of the larger particles, often above the nanoparticle range, being most evident. In a similar way, any attempt to investigate their structure by conventional diffraction methods produces a further average result, and obtaining even this is fraught with problems when allowance has to be made for effects such as diffraction line broadening (Stokes, 1948) arising from their very small size. Normally this is virtually impossible if the overall size distribution is not known, or if strain or relaxation effects are present in individual particles. Consequently, all that can normally be done is to relate whatever structural features which are displayed in the diffraction pattern to those of the bulk material: any differences which might exist are generally lost in the background, or any visible discrepancies are very difficult to explain unless some form of modelling is attempted.

Particulate Matter, edited by R.L. Maynard and C.V. Howard

It is only in the last 20 years, with the advent of high resolution electron microscopy (HREM) at sub-0.2 nm levels, and the consequent ability to resolve inter-atomic spacings at this level, that any real attempt has been made to determine the atomic structure of *individual* particles. All the studies described in this chapter rely heavily on HREM images, but although often appearing very spectacular, their information content should not be over-estimated. HREM avoids the phase problem which is associated with diffraction methods, but this is replaced by an 'instrumental phase problem' (Scherzer, 1949) arising from the spherical aberration of the electromagnetic lenses used in the image formation process. In a diffraction experiment the overall resolution (the diffraction limit) is equal to one-half of the wavelength of the radiation employed:for Mo Kα radiation this corresponds to 0.0355 nm, which is much less than an atomic diameter (ca. 0.1–0.25 nm). In a modern high resolution electron microscope, the effects of spherical and chromatic aberration, coupled with the inherent energy spread of the electrons from a heated cathode (Hanzen, 1971) and the finite divergence of the electron beam (Frank, 1973), combine to limit the resolution to at best 0.12–0.13 nm. This is a much poorer figure, and consequently much of the information which is taken for granted in a diffraction study, such as accurate bond distances, is not available in a HREM study. For this reason, even the structural information concerning nanoparticles which is now becoming available is comparatively crude.

5.1.1 The role of surface atoms

The feature which makes nanoparticles so interesting is the relatively high proportion of surface atoms which they possess. Consequently, one might expect that particles should therefore adopt a spherical morphology to minimize their surface area and hence their overall energy. For spherical particles of average radius r_{av}, the surface area per unit mass, σ, may be calculated approximately by:

$$\sigma = 3/(\rho r_{av}),$$

where ρ is the bulk density, and if r_a is the mean atomic radius, the ratio of the number of surface atoms to total atoms, F_s may be roughly expressed as:

$$F_s = 4r_a/r_{av}.$$

Both of these quantities increase as the mean particle radius r_{av} decreases, such that an assemblage of particles with a mean diameter of 5×10^{-9} m has almost one atom in six on the surface, and for a typical density of about 50 kg m^{-3}, a value of σ of approximately 2.4×10^5 m^2 kg^{-1}. Such a high proportion of surface atoms ensures that, in general terms, nanoparticles of this size regime display vastly increased reactivity, which explains their numerous applications in heterogeneous catalysis. This reactivity, in turn, can have far-reaching consequences, as can be shown by the surface modification of certain metal particles. Not the least of these consequences is the role that the surface atoms may thus play in determining the structure of the *entire* particle, and not just the surface regions.

5.1.2 The effect of surface reactivity

One of the most striking examples of the latter behaviour concerns the poisoning of platinum catalysts by sulphur. When used in catalytic applications, platinum is normally dispersed in the form of metal nanoparticles on an inert support such as α-alumina (prepared either by colloidal routes or from a chloroplatinate salt by reduction in H_2) with sizes in the range 1–10 nm. Unlike the particles of silver and gold described below, these particles are roughly spherical in form, but it has long been known that in the presence of sulphur this morphology alters to that of small cubes (Harris, 1986) with the same overall size, and the particles become catalytically inactive. Closer examination of these cubic particles however (Jefferson and Harris, 1988; Gribelyuk *et al.*, 1994), reveals that at their surfaces the periodicity of the bulk structure is doubled and the faces parallel to the electron beam take on a 'castellated' appearance (*Figure 5.1(a)*). Using image simulation techniques based on the multislice algorithm developed for multiple scattering of electrons (Cowley and Moodie, 1957, Goodman and Moodie, 1974) from models of complete particles (Uppenbrink *et al.*, 1990) it may be shown that this distinctive surface structure arises from a layer of sulphur atoms adsorbed on the four-fold symmetry sites of the so-called "cube" faces of the structure of platinum, the final model being shown in *Figure 5.1(b)*. This $c(2 \times 2)$ overlayer of sulphur on the platinum cube faces is extremely stable, and the large enthalpy of adsorption, ΔH_{ads}, evidently provides the driving force not only for the reaction with sulphur itself but also for the complete re-arrangement of the particles from a spherical morphology to a cubic one, thereby maximising the area of the these cube faces on which the adsorption can occur. Although metal atoms are relatively mobile at elevated temperatures, the energies involved are by no means negligible, as may be seen in the case of platinum–palladium alloys, which on sulphiding (Narain, 1995) separate into cubic platinum particles, with their adsorbed overlayer, and palladium particles which do not undergo such a dramatic surface reaction (*Figure 5.1(c)*). The free energy of mixing of an ideal 50:50 alloy solid solution at a typical sulphiding temperature (773 K) is approximately 5.4 kJ mole^{-1} so the stabilization energy of the sulphur overlayer must be of this order of magnitude at least to make the resulting exsolution process possible.

5.2 Structures of metal nanoparticles

The fact that surface energy terms in nanoparticles, such as the ΔH_{ads} for sulphur on platinum, are comparable with bulk energies, has very

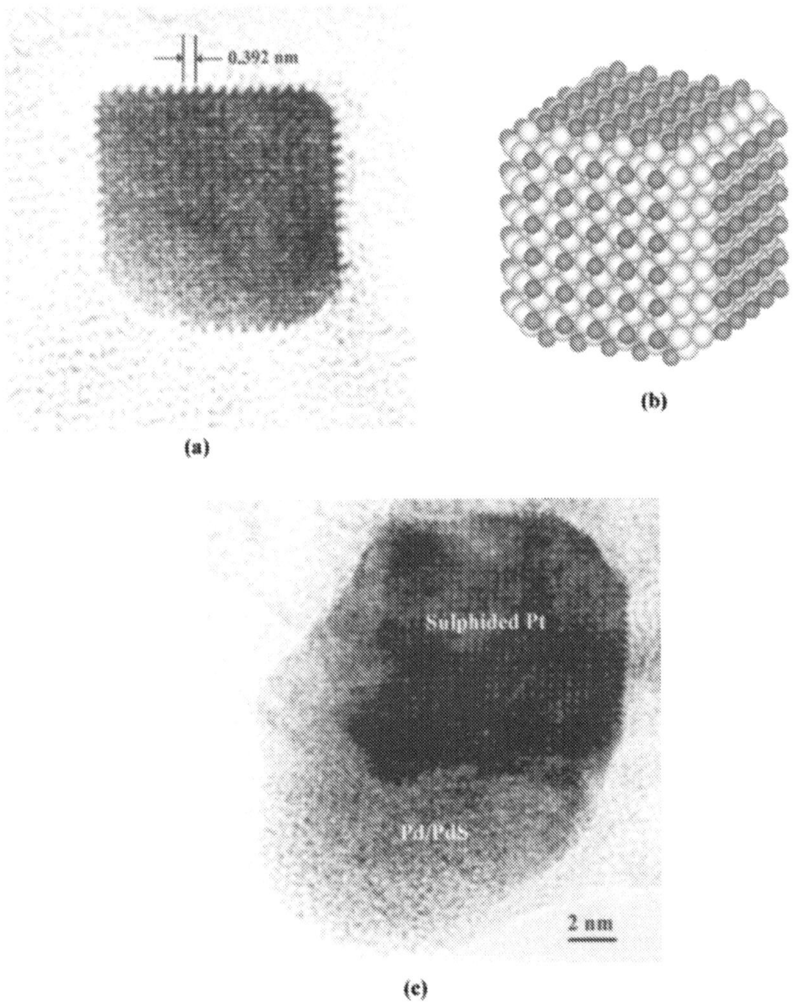

Figure 5.1. (a) HREM of a cubic nanoparticle of platinum resulting from the formation of an ordered sulphur overlayer. (b) Schematic model of such a particle, showing sulphur atoms adsorbed on the four-fold surface sites. (c) HREM of a 50/50 platinum/palladium alloy nanoparticle, after reaction with sulphur. The separation of a sulphided platinum particle is clearly visible.

important consequences in terms of the overall particle structure. Until quite recently, it has invariably been assumed that the structures of nanoparticles were identical to those of the equivalent bulk materials and that their surfaces were therefore merely terminations of the normal bulk structure. If this were to be the case, then properties dependent on surface area, such as catalytic activity, should show a clear $1/r$ dependence on r. However, this is frequently not the case, and often it is found that the reactivity follows a $1/r$ dependence down to a certain critical value, and then either increases much more rapidly or alternatively, starts to decrease. The inference is that when particles enter the nanoscale range, the nature of the surface may change, and this may in turn result in alterations to the structure of the interior.

The simplest nanoparticles are those of metals. Most metals can be regarded as structures built up by the close-packing of atoms which may be considered as hard spheres. In a 2-dimensional close-packed layer, each metal atom is in contact with six nearest neighbours. Successive layers in the structure cannot superimpose immediately above one another, as the resulting structures would not be close-packed, but instead the atoms of one layer fit into the indentations between atoms of the layer below. Two configurations are possible, one, called hexagonal close-packing of hcp, where the third layer is located exactly above the first, and cubic close-packing or ccp, where the sequence repeats with the fourth layer. In both each metal atom is 12-co-ordinated by other atoms, i.e. each has 12 nearest neighbours. Cubic close-packing is by far the more common, and the arrangement of atoms is such that they occupy the corners and face-centres of a cubic basic unit or unit cell. For this reason the arrangement is often referred to as face-centred cubic or fcc. Crystallographically, planes within a structure are described in terms of Miller Indices, these being integers such that the intercepts of a plane (hkl) on the crystallographic axes x, y and z are a/h, b/k and c/l, where a, b and c are the repeat units along the respective axes. In the fcc structure the close-packed plane is then the (111) plane, and the cube face, referred to above, is the (100) plane. As there are more than one equivalent planes of this type within each unit cell these are normally referred to as planes of the family {111}, etc. In metal nanoparticles the surface planes or facets exhibited are normally designated by these Miller indices.

5.2.1 Surface facetting

The classic example of this is found in the case of the coinage metals, especially silver and gold. Once again the dominant force is the free energy of the particle surfaces. In a truly spherical particle of an fcc metal (illustrated schematically in *Figure 5.2(a)*) many of the atoms are situated in close-packed planes with a co-ordination number of 9, but there are a substantial number sited at surface steps with a lower than average co-ordination number. These are inevitably of high-energy (although they are the sites which are commonly regarded as being catalytically active) but if a true spherical morphology is adopted they are inevitable. However, the proportion of these sites will rise with decreasing r_{av}, and if the energy is to be continually minimised, their presence is unacceptable and some other solution must be found. The most common way in which fcc metals accommodate this problem without alteration of the internal structure is by *facetting* (Marks and Smith, 1981). This entails the abandonment of a truly spherical morphology and its replacement by a configuration which is only approximately spherical, but displays regular crystallographic faces having a relatively high

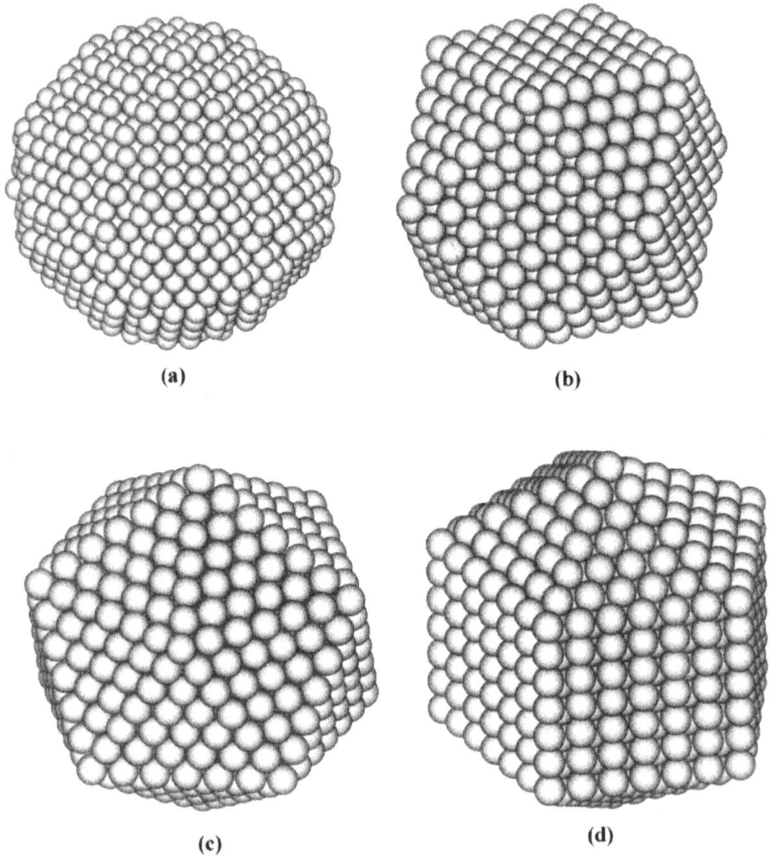

Figure 5.2. (a) Schematic model of a spherical metal nanoparticle with the fcc structure, showing atoms at surface steps. (b) A model of a cubeoctahedral particle of similar size, containing 777 atoms. (c) Icosahedron with the same number of atoms, constructed using the homogeneous strain model. (d) decahedron, also with 777 atoms.

surface co-ordination number. The most common solution adopted in fcc metals is therefore the cubeoctahedron, illustrated schematically in *Figure 5.2(b)*. This comprises eight {111}-type faces, with a surface co-ordination number of 9, linked by six {100}-type faces, where the equivalent number is 8. A particle of this type has a greater proportion of surface atoms than a true sphere containing the same number of atoms, but the average surface co-ordination number, and hence the overall surface energy, is nevertheless reduced. The example shown in *Figure 5.2(b)* is perfectly regular, with the {111}-type faces being triangular, but this need not be the case, and extending these faces at the expense of those of the {100}-type will raise the average surface co-ordination number, albeit at the cost of also increasing the number of surface atoms. The dominating factor is inevitably the relative energies of bulk and surface atoms.

5.2.2 Non close-packed nanoparticles

Although the cubeoctahedron and related morphologies provide a partial solution to the problem, there are other ways of increasing the surface co-ordination number by facetting while preserving a near-spherical morphology, and the most interesting of these is the icosahedron, illustrated schematically in *Figure 5.2(c)*. This is best imagined as consisting of 20 tetrahedral elements placed together such that one close-packed face is outermost, any given tetrahedron being related to its neighbours by twinning on the other three close-packed faces. All the tetrahedral elements therefore share one common vertex at the centre of the particle. Twinning on the close-packed {111} planes is common in fcc metals, the angle between close-packed rows of atoms in adjacent twins being 70.52°. In an icosahedron, however, this angle must be 72°, as the tetrahedral elements are arranged in groups of five, so some distortion of the structure is necessary. This distortion is manifest as an elongation of the close-packed rows of atoms parallel to the outer faces of the icosahedron, with a corresponding relaxation of the atoms in a radial direction towards the centre of the particle. This is the so-called 'homogeneous strain model' (Howie and Marks, 1984). The net result, therefore, is that the structure is no longer close-packed, and consequently the volume free energy must rise, but this increase is compensated for by the reduction in the surface free energy arising from the exclusive adoption of (approximately) close-packed surfaces, with their average co-ordination number of 9.

Icosahedral particles are not the only non close-packed solution possible, and the possibility of producing a decahedral morphology also exists, this being illustrated schematically in *Figure 5.2(d)*. In this, the particle contains only five individual elements, arranged around a unique axis, but still subject to similar structural distortions as those found in the individual components of icosahedral particles. Because the number of individual compenents is reduced the interfacial energy associated with their presence will also be lowered. However, if only the close-packed surfaces are favoured the decahedron is a relatively flat particle, with a consequently high fraction of surface atoms, but this fraction can be lowered by introducing {100}-type facets parallel to the five-fold axis, as shown in the case of *Figure 5.2(d)*. As in the cubeoctahedron, these faces have a lower co-ordination number (8), but there are fewer of them than in a cubeoctahedron containing the same number of atoms, and the energy is therefore lowered. Icosahedral and decahedral morphologies are normally grouped together and referred to as multiply twinned particles or MTP's (Marks, 1983). In metals such as silver and gold they are extremely common. Decahedra are readily recognisable if viewed down the five-fold symmetry axis, as the metal atoms then superimpose in projection, as shown in *Figure 5.3(a)* (Smith and Marks, 1981). Icosahedra, which display five-, three- and two-fold symmetry axes, on the other hand, are much more difficult to recognise. When viewed down the two-fold axes atoms at the edges will project on top of one another and can

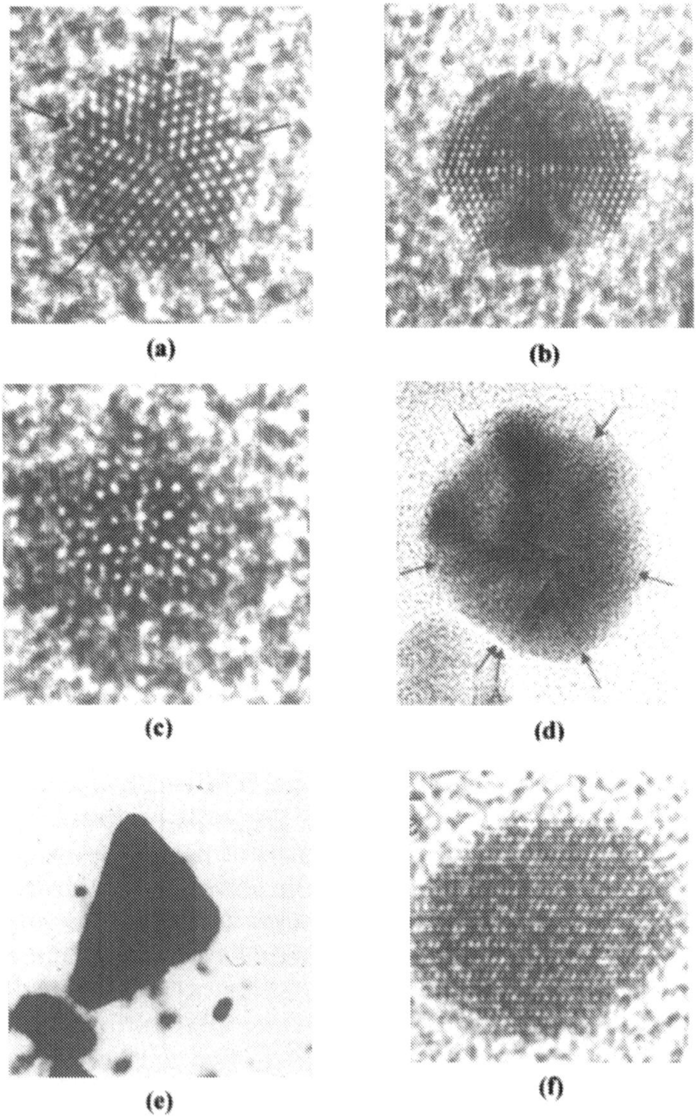

Figure 5.3. (a) HREM of a decahedral particle of silver, viewed down the five-fold symmetry axis. (b) HREM of icosahedral silver, viewed down one of the two-fold axes. (c) A smaller icosahedron of silver, viewed down one of the five-fold axes. (d) A larger decahedron, also of silver, showing the introduction of a defect strip (A) caused by excessive strain in the atomic arrangement. (e) A "trigon" of colloidal gold, viewed perpendicular to the plane of the particle. (f) HREM of a nanoparticle of colloidal copper, showing a non-standard stacking sequence of layers.

therefore be imaged quite readily, as shown in *Figure 5.3(b)* (Marks and Smith, 1983), but when viewed down the five-fold axes none of the atoms superimpose : the corresponding images, such as that shown in *Figure 5.3(c)* and confirmed by image simulation methods (Jefferson and Kirkland, 1988; Kirkland *et al.*, 1991) using the periodic object approximation

(Grinton and Cowley, 1971), are therefore much more difficult to interpret even at the 0.12–0.13 nm resolution limit currently available in HREM, despite the fact that they show considerable contrast.

5.2.3 The role of kinetics

Because they display five-fold symmetry, both icosahedral and decahedral nanoparticles have structures which are incompatible with a periodic lattice. It is therefore incorrect to describe them as small crystals, and as the environment of each atom is not identical, their overall properties can only be the average over all the atoms present. As their size increases, then the elongation of the close-packed rows parallel to the particle surfaces increases accordingly, and at a certain stage the total elongation is sufficient to accommodate an extra atom and defects may be introduced. This is the case in the large particle shown in *Figure 5.3(d)*, where the overall morphology is basically still that of a decahedron, but an extra segment is introduced to take up the misfit between the original five elements, and the bulk of the particle presumably reverts back to the normal close-packed structure. At this point the extra energy associated with the defect must be outweighed by the decrease in energy associated with the change from the ideal non close-packed decahedral atomic arrangement to the normal close-packed fcc structure. However, we must be careful in discussing such nanoparticles only in terms of energies: if energy were the only consideration then decahedral particles would not exist (Uppenbrink *et al.*, 1992), for they invariably possess higher energy than their icosahedral counterparts with the same number of atoms. The only explanation for their formation is that they result from kinetic, rather than thermodynamic factors, and once formed, continue growing in their distinctive morphology. Further evidence for the role of kinetics in determining nanoparticle structures comes from the presence of larger, flat nanoparticles in colloidal preparations of gold and silver where reduction of the metal salt has proceeded relatively slowly (Heinemann *et al.*, 1979, Kirkland *et al.*, 1993). These particles, called trigons, have a structure which is basically close-packed but elongated parallel to one of the {111} planes of the fcc arrangement. One such particle is shown in *Figure 5.3(e)*. The reason for their apparent anisotropy appears to be the presence of one or more stacking defects at their centres, which then provide sites for further growth in a lateral direction. In nanoparticles of other metals even more bizarre phenomena are observed. For example, platinum does not form MTP's but nanoparticles of this metal are normally approximately spherical in shape with (presumably) the fcc structure, although they do contain large numbers of stacking faults:copper, on the other hand, appears to favour much more ordered sequences of layers which are not found in either hexagonal or cubic close-packing (*Figure 5.3(f)*). This suggests that considerations such as the stacking fault energy and even the basic energy difference between hexagonal and cubic close-packing may well be a function of particle size.

5.3 Metal oxide nanoparticles

Metal oxide nanoparticles have not been studied as extensively as those of pure metals, largely because they are more difficult to prepare and to characterize using diffraction and HREM techniques. In particular, metal-oxygen distances are of the order of 0.1–0.15 nm and still cannot be resolved in most microscopes, with the result that only the metal atoms may be imaged. However, insofar as their properties are concerned, the same factors are involved, namely the balance between surface and bulk free energy. Consequently, most oxide nanoparticles are found to eliminate surface steps and display crystallographic faces with varying degrees of perfection, the latter tending to increase as the particle size is reduced below 10 nm. However, in oxides, unlike the case of metallic particles, a second most important factor arises, namely that of maintaining chemical stoichiometry in a particle of finite size. Although some variation is possible if the metal atom is multivalent, as in transition metal oxides, in binary oxides such as Al_2O_3 there can be no compromise and the chemical formula must be adhered to. As we shall see, this poses extreme problems in terms of structure, but various solutions are possible.

5.3.1 Bulk-terminated nanoparticles – TiO_2

An excellent example of the consequences of maintaining low-energy surfaces in an oxide nanoparticle is illustrated by the anatase form of TiO_2, which is frequently used as an 'active' catalyst support. *Figure 5.4(a)* shows a micrograph of an anatase nanoparticle, prepared colloidally for use as a support in the electron microscope (Harris *et al.,* 1983), and viewed down a [110]-type axis. The structure is tetragonal, and the terminating faces of the crystal are of the {001}- and {011}-type. The image of *Figure 5.4(a)* reveals clearly the cation arrangement, but the oxygen atoms are not resolved. The particle surfaces are well-defined, but their structure is not obvious. Anatase is not a close-packed structure, but the atomic arrangement is similar on the {001} and {011} planes, consisting of planes containing both titanium and oxygen atoms interspersed with planes containing only oxygen. Bearing in mind the reactivity of titanium, conventional chemistry might predict a surface terminating only in oxygen atoms, but this would inevitably lead to an oxygen excess in the crystal. Consequently this apparent excess must be countered either by introducing oxygen vacancies in the interior of the particle, which are known to exist in bulk material, or the surface of the particle must involve termination in both titanium and oxygen atoms. From the characterization viewpoint, the oxygen atoms cannot be resolved, but their presence, or otherwise, at the particle surface may be deduced from their effect on the structure of the Fresnel fringe existing

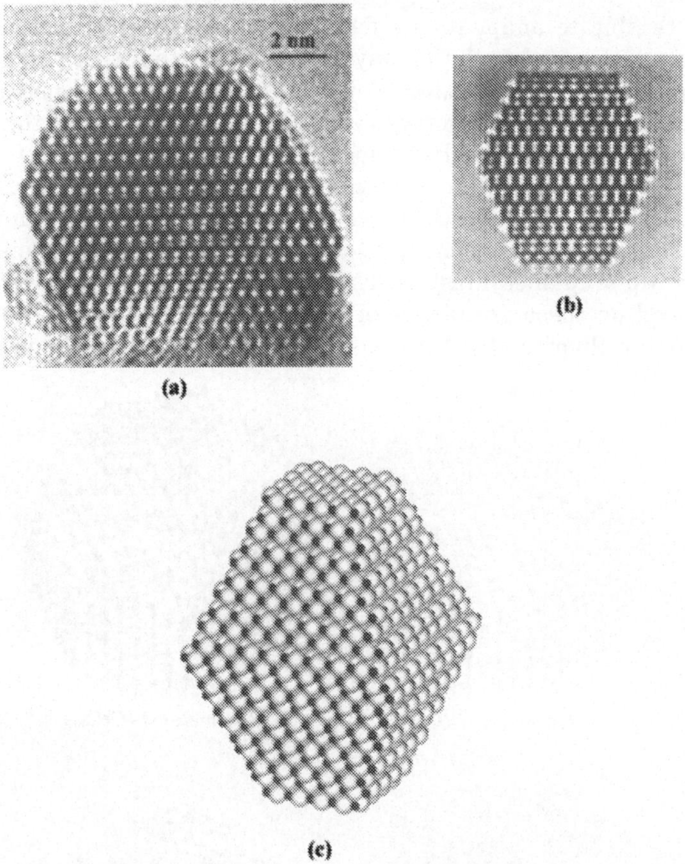

Figure 5.4. (a) HREM of a nanoparticle of anatase (TiO_2), viewed down the [110] axis. Spaces between the titanium atoms are revealed as white dots. (b) Simulated images of a complete particle of this type with surfaces terminated in planes containing titanium and oxygen atoms, reproducing the observed edge contrast. (c) Model of the latter, with surface titanium atoms in five co-ordination.

in the image at the margins of the particle. This fringe, which corresponds to the sharp drop-off in potential at the edge of the particle, is extremely sensitive to changes in the potential gradient, which is much less steep if terminating oxygen atoms are present. Image simulation studies using model particles of the type shown in *Figure 5.4b*, with different types of surface termination arrangement (Kirkland *et al.*, 1991), indicate that both the {001} and {011} surfaces do not consist solely of oxygen atoms, but also contain titanium. Although this is not totally unexpected, it does imply the presence of titanium in a distorted five-fold co-ordination environment, suggesting that the reactivity of these atoms may be higher than those in the interior. The final model used for the anatase nanoparticle is illustrated in *Figure 5.4(c)*.

5.3.2 Nanoparticles with anion vacancies – CeO_2

Anatase is able to adapt to the formation of particles of very limited dimensions more readily than many oxides, and with little change to the basic oxide structure. Other oxides, however, have a much more fundamental stoichiometry problem which may be illustrated by considering the model cubeoctahedra of CeO_2 which are shown in *Figure 5.5(a)*. CeO_2 is readily prepared in nanoparticlulate form and is thus particularly susceptible to this problem (Brinker and Scherer, 1990). To minimize the number of surface steps and consequently the overall surface energy, low-index crystallographic planes must be used for the terminating faces:in the cubeoctahedron these are planes of the {111}- and {100}-type. However, CeO_2 has the fluorite structure, and the arrangement of atoms in this

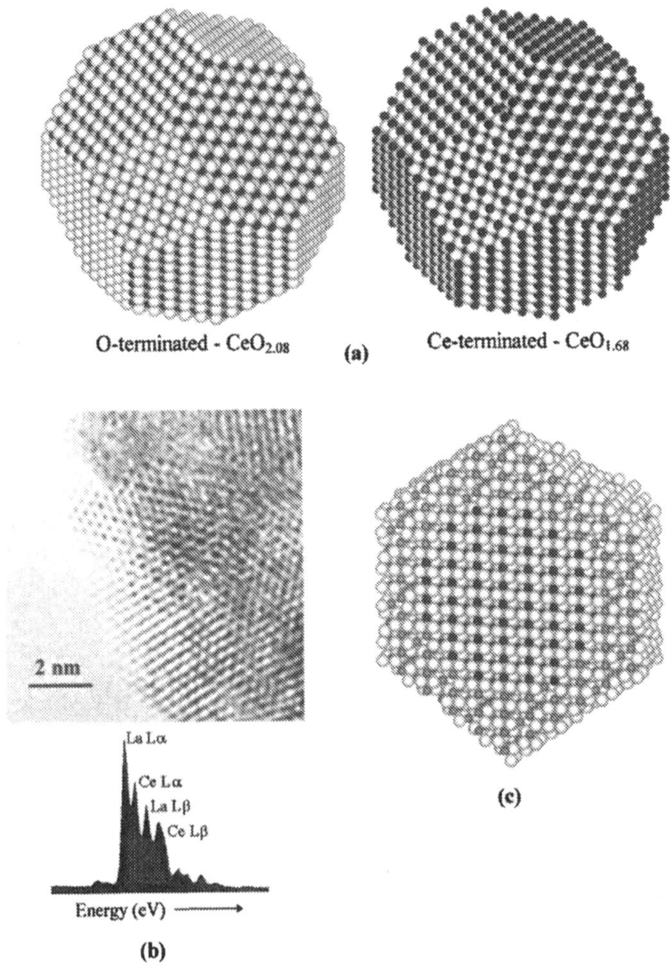

Figure 5.5. (a) Models of cubeoctahedral nanoparticles of CeO_2 illustrating the stoichiometry problem arising from cation and anion termination. (b) A micrograph of a La_2O_3-CeO_2 nanoparticle, with x-ray emission analysis showing the presence of both oxides. (c) A model of the proposed structure, with and external 'shell' of C-La_2O_3 around a CeO_2 core.

structure is such that individual {111} or {100} planes may contain either cations or anions, but not both simultaneously. Consequently, a nanoparticle of CeO_2 must have an excess of either anions or cations depending on which of these ions constitute the terminating planes of the particle. For sub-5 nm particles this problem becomes acute, as shown by the models illustrated. The only way in which this stoichiometry problem may be solved is via the creation of vacancies, either cation vacancies in the case of cation-terminated surfaces, or anion vacancies if the surfaces have anions as the outermost components. In view of the number of anion-defective fluorite structures known, particularly at high temperatures (Brauer, 1964; Bevan, 1973), the latter seems far more probable.

If anion vacancies are present in nanoparticulate CeO_2, their characterization presents problems, as the oxygen atoms are not visible in HREM images at current levels of resolution, and simple image simulation experiments show quite readily that the contribution of these atoms to the image contrast is almost negligible. Some indications as to the nature of the vacancy arrangement, however, come from indirect evidence, particularly the solid solubility of other oxides in CeO_2 nanoparticles. Solid solutions of La_2O_3 and CeO_2 are a particularly striking case, as the C-form of La_2O_3 has a structure which is based on the fluorite type, but with ordered oxygen vacancies to permit the accommodation of a cation (La^{3+}) with a lower nominal valence, this arrangement necessitating a doubling of the unit cell edges (Gschnieder and Eyring, 1979). Owing to the similarities in structure, the determined maximum solid solubility of La_2O_3 in bulk CeO_2 is found to be quite high at elevated temperatures, being approximately 50% at 1500 K and over 60% at 1900 K (Bevan, 1955; Morris *et al.*, 1993). If the CeO_2 is in nanoparticle form, however, the low temperature solubility is very markedly increased, up to a maximum of 63%, even in specimens prepared by 'wet' methods which are annealed at no more than 700 K (Tilley, 1997). A nanoparticle of this type is shown in *Figure 5.5(b)*. Furthermore, there is no tendency for an increase in particle size as the proportion of La_2O_3 is increased, despite the fact that it is normally much more difficult to prepare sub-10 nm particles of La_2O_3 than CeO_2. The inference must therefore be that this excess presence of La^{3+} cations, together with their associated oxygen vacancies, is stabilized due to the nanoparticle nature of the host CeO_2 and the consequent need to prevent an excess of anions in the particle. Although the anions cannot be observed using HREM, and the lanthanum and cerium atoms cannot be distinguished, the overall composition of these nanoparticles may be measured using X-ray emission analysis in the electron microscope, and the 63% maximum solubility confirmed. However, if a surface-sensitive technique such as X-ray Photoelectron Spectroscopy (XPS) is used the apparent lanthanum content is much higher, inferring that the surface is rich in lanthanum. If this is the case, then the anion vacancies must also be located in the surface region, and the nanoparticle structure will resemble the schematic model of *Figure 5.5(c)*, where a central 'core' of stoichiometric

CeO_2 is surrounded by a 'skin' of C-La_2O_3. In the light of these findings, it seems highly probable that nanoparticles of pure CeO_2 have a similar structure, and the presence of surface anion vacancies would explain the apparent ease with which such nanoparticles can take up and release oxygen in their role as oxygen storage devices (Perrichon *et al.*, 1994).

5.3.3 Spinel oxides and cation vacancies – Fe_3O_4

Oxides based on the rock-salt or spinel structures also suffer from the same problem as CeO_2, namely that there are no low-index crystallographic planes which contain both anions and cations. Materials with these structures generally form octahedral nanoparticles terminated in {111} planes, with the apices frequently blunted by {100} faces to form distorted cubeoctahedra. One of the most striking illustrations of the effect of the stoichiometry problem in this type of oxide is seen in the iron oxides. $Fe_{1-x}O$ is the simplest oxide, adopting the rock-salt structure with all Fe^{2+} ions in octahedral co-ordination, but containing also some Fe^{3+} ions, the presence of which gives rise to cation vacancies in the structure (Reed, 1970). The trivalent Fe ions appear to be placed on both the normal octahedral sites and tetrahedral (interstitial) positions, but there is no firm evidence to date of any ordering of the metal vacancies. If the number of Fe^{3+} ions is increased significantly, however, then ordering takes place and the Fe_3O_4 structure forms. This is a spinel oxide, with the same fcc array of oxygen anions as is found in rocksalt, but with Fe^{2+} and Fe^{3+} ions ordered onto both octahedral and tetrahedral sites which form interpenetrant chains within the structure. The size of the unit cell also doubles when this ordering takes place. Further oxidation of the remaining Fe^{2+} ions then produces Fe_2O_3, which may adopt two very different structures. The first of these is the haematite structure, which has a hexagonal close-packed array of oxygen anions and is closely related to that of corundum, α-Al_2O_3, but a second variant, γ-Fe_2O_3 or maghemite (Sinha and Sinha, 1958), retains the spinel structure but distributes, on average, 21.3 Fe^{3+} cations over 24 structural sites.

Under normal conditions, all three phases are thought to possess their own stability fields, governed by the temperature and the partial pressure of oxygen. However, at low temperatures, it may be postulated from thermodynamic criteria that the cation-deficient $Fe_{1-x}O$ should undergo a phase separation into stoichiometric FeO and Fe_3O_4 according to the equation:

$$n.Fe_{1-x}O \rightarrow n(1\text{-}4x).FeO + nx.Fe_3O_4$$

This reaction cannot normally be investigated in bulk material as the temperature required is too low for effective diffusion of Fe atoms, but, if microcrystalline $Fe_{1-x}O$ is used as the starting material, it may be observed in the electron microscope under gentle heating in the electron beam, when the margins of crystals of $Fe_{1-x}O$ become expitaxially coated with small crystallites of Fe_3O_4, which are easily recognisable by their doubled periodicity. This reaction can be proved to be a disproportionation process, rather than in situ reduction, as increasing the beam current raises the temperature sufficiently for the process to be reversed and the equilibrium shifted back to the left (Jefferson *et al.*, 1992). Although the reaction principally involves motion of the Fe atoms, there is some recrystallization at the surface due to movement of oxygen atoms also and much of the Fe_3O_4 produced is in nanoparticle form.

A micrograph of one such nanoparticle is shown in *Figure 5.6a*. This particle has grown epitaxially on the original crystal, and the surface

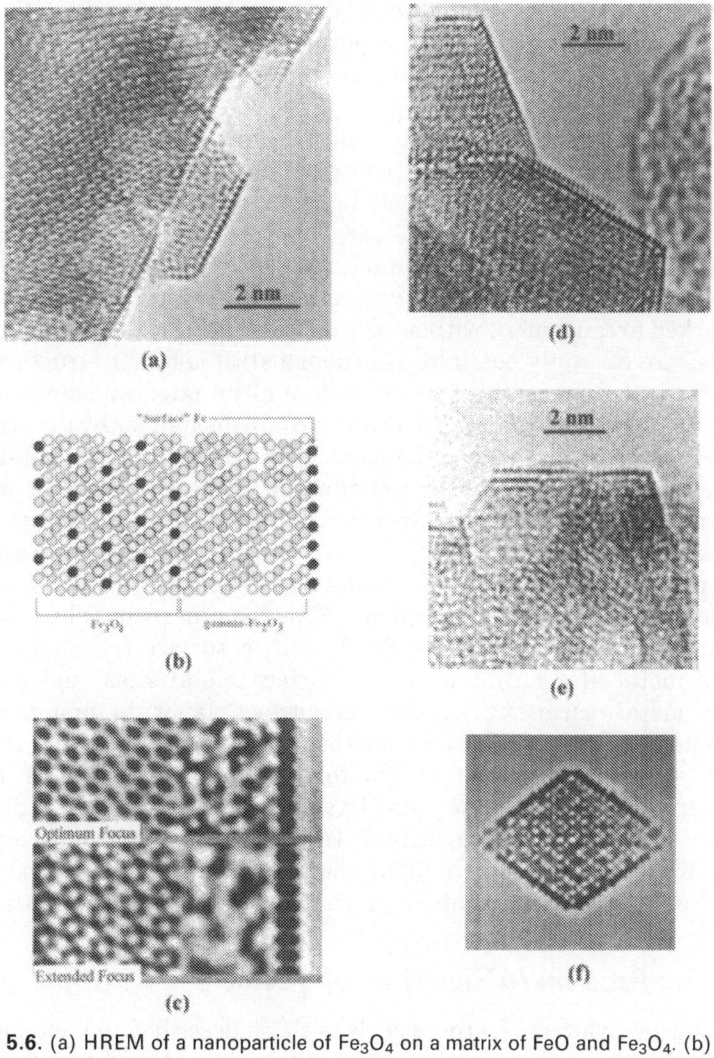

Figure 5.6. (a) HREM of a nanoparticle of Fe_3O_4 on a matrix of FeO and Fe_3O_4. (b) A model for the proposed structure, with continuous anion lattice. (c) Images simulated from this model. Reproducing the contrast at the surface layer. (d) and (e) large and small particles of γ-Al_2O_3 respectively showing similar surface contrast. (f) Simulated image from a model nanoparticle, again reproducing the surface contrast.

layer of the latter has also been converted to Fe_3O_4. The nanoparticle is apparently bounded by {111} and {100} planes, and almost certainly has a cubeoctahedral morphology. The most interesting aspect of this particle, however, is the structure revealed in the HREM image. In this [110] projection, the cation-cation distances within Fe_3O_4 are easily resolvable and clear lattice fringes should be visible throughout the particle, although any contribution from the anion lattice is extremely weak by comparison. Nevertheless, this is not the case in the actual image, where, although the individual cations are clearly outlined at the particle edges, with very high contrast, there is little if any regular contrast from the interior of the particle, the overall pattern there being typical of an amorphous object. This is clearly a paradoxical situation, as there is no reason why an amorphous oxide should possess faces typical of a crystalline structure.

This paradox may be resolved if the stoichiometry problem faced by nanoparticles of Fe_3O_4, and the contribution of both iron and oxygen atoms to the final image contrast, is considered. The {111} and {100} planes which form the cubeoctahedral morphology can contain either cations or anion, but not both simultaneously. Unlike the case of CeO_2 there is no way in which anion vacancies can be incorporated into the close-packed arrangement without breakdown of the structure, but cation vacancies are certainly possible, as is demonstrated by the structure of γ-Fe_2O_3. If these nanoparticles are therefore terminated by planes of iron atoms, the cation excess so caused can be compensated for by the creation of numerous vacancies within the particle interior. The anion sub-lattice remains intact, but because the contribution of the anions to the overall image contrast is so small, this regular component of the structure is not observed in the image, and all that can be seen is the contrast resulting from a random arrangement of cations and cation vacancies, which will consequently appear to be amorphous. This hypothesis may be tested by constructing a model of regular Fe_3O_4 with a surface terminating in a plane of metal atoms, filling all the surface metal sites, and creating random metal vacancies in the sub-surface layer to maintain the stoichiometry. The anion lattice remains continuous throughout, and this model is shown in *Figure 5.6(b)*. The images simulated from this (*Figure 5.6(c)*) reproduce the experimental image contrast very well, indicating the basic soundness of the principal. To obtain sufficient contrast at the surface layer, it is necessary to fill all the octahedrally co-ordinated sites in the surface layer with metal atoms, as in the manner of stoichiometric FeO.

5.3.4 Surface metallization of γ-Al_2O_3

Strong surface contrast features such as those described in nanoparticles of Fe_3O_4 have also been observed in other spinel-based oxides. Beam induced re-crystallization of α-Al_2O_3 into the γ-form, which has the defect spinel structure, has also been noted (Smith *et al.*, 1986). γ-Al_2O_3 is difficult to prepare in the pure state, and must normally be examined as part of a mixture with the α-form, but there is strong thermodynamic evidence that, as the particle size decreases, the former becomes the thermodynamically stable structure (McHale *et al.*, 1997). Al_2O_3 is widely used as a support for metal catalysts, and there has long been some suspicion that the so-called 'active' support, which facilitates the monodispersion of metals, is in fact the γ-form.

When nanoparticles of γ-Al_2O_3 are examined by HREM the images show strong similarities to those of equivalent particles of Fe_3O_4, as shown in the image of *Figure 5.6(d)*. In the image shown the particles are at the higher end of the nanoparticle size regime, but they still show the strong contrast at their edges, with only weak fringe contrast in the interior regions. The only notable difference from the Fe_3O_4 nanoparticles is the severe truncation of the {100}-type faces so that the overall particle shape is octahedral. Although the fringe contrast of the central regions of larger particles such as those in *Figure 5.6(d)* is weak but present, in the smaller particles it is virtually absent altogether, and as is the case with the Fe_3O_4 particles described above, the interior appears to be amorphous. *Figure 5.6(e)* shows such a particle. The images of these nanoparticles can be explained using exactly the same model as with those of Fe_3O_4, namely particles having surfaces which terminate in metal atoms and an excessive number of cation vacancies in the interior. This is shown very clearly in the image which is simulated from a model particle and is shown in *Figure 5.6(f)*. As the particle size increases, the proportion of vacant sites in the interior is reduced and the conventional lattice fringe contrast starts to appear, but only weakly. Furthermore, because the difference in the scattering power of oxygen and aluminium is less than that between oxygen and iron, there is no doubt in this material that the surface is metal-terminated, as no other model can explain the pronounced contrast at the particle edges. Bearing in mind the reactivity of aluminium, this is chemically very surprising, but it may explain the apparent ease with which metals such as platinum and rhenium disperse when supported on γ-Al_2O_3. It seems highly likely that, when these metals are added to a specimen made up of γ-Al_2O_3 nanoparticles, they actually 'dissolve' in the surface metal layer and thus release aluminium ions which migrate to the particle interior, further stabilizing the particle. The exact valence of metal atoms added in this way has not yet been determined, but it is unlikely to be zero, explaining the extreme reactivity of such catalysts.

5.3.5 *New structure types in* WO_3

All the nanoparticle structures described so far, although unique to the particles concerned, are basically modifications of structures found in bulk material. One of the intriguing possibilities in nanoparticles, however, is the formation of new structures which have no bulk counterpart, and one such structure has been obtained in nanoparticles of tungsten trioxide. There are three known structures for tungsten trioxide, one of which is a distorted form of the simple perovskite structure (ABO_3) with the A-cation removed and only the network of corner-sharing WO_6 octahedra remaining. This structure, which is shown in *Figure 5.7(a)*, is normally referred to as the monoclinic form, m-WO_3 (Wells, 1984), although it becomes cubic at temperatures much above 1200 K. There are two further structures which have been prepared using 'wet' methods, namely a hexagonal form, h_1-WO_3, and a pyrochlore-like form, p-WO_3 (Figlarz, 1989). The hexagonal form illustrated in *Figure 5.7(b)*, which contains tunnels formed by six corner-sharing WO_6 octahedra, is prepared by dehydrating the hydrated tungsten oxide WO_3. ⅓ H_2O but converts irreversibly to m-WO_3 at temperatures above 700 K (Gerand *et al.*, 1979). The pyrochlore form is prepared from tungstates of Cs^+, RB^+ or NH_4^+, which themselves have a pyrochlore structure, by removing the cation via treatment with HNO_3. The structure of the pyrochlore form contains tunnels similar to those in h_1-WO_3 but they are no longer straight and exist in all three dimensions. The hexagonal form is basically a pure oxide equivalent of certain alkali tungsten bronzes (Ekstrom and Tilley, 1980). It is not certain whether h_1-WO_3 or p-WO_3 are really thermodynamically stable.

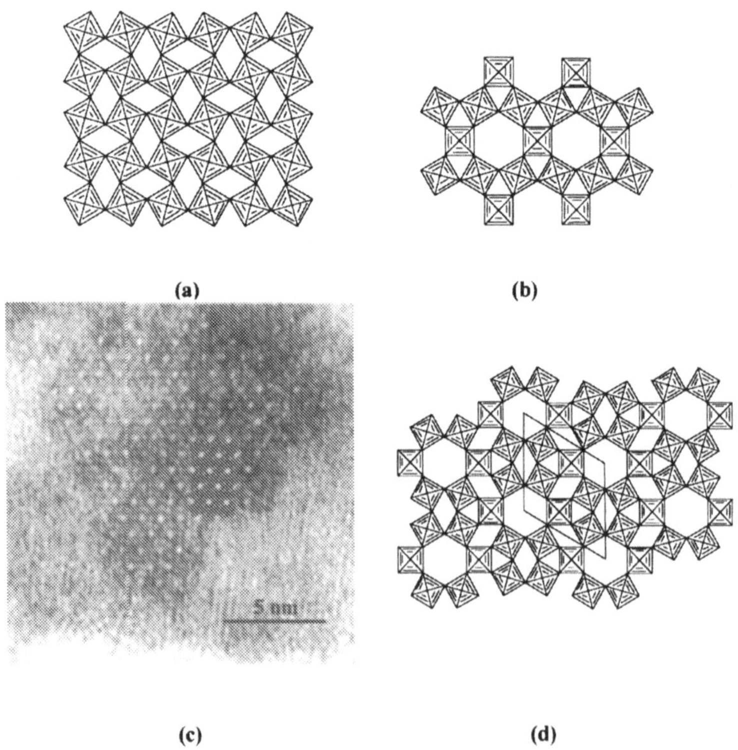

Figure 5.7. (a) and (b) The structures of m-WO_3 and h_1-WO_3. (c) HREM of the new structure, only observed in nanoparticle form. A simulated image is shown inset (d) Schematic representation of the new structure.

Nanoparticles of WO_3 may be prepared very effectively using sol-gel techniques and employing acidified sodium tungstate as the starting material, the resulting gel being refluxed with either 30% H_2O_2 or NH_4Cl solution at a higher pH until a fine yellow precipitate is formed. When examined using HREM, the final product invariably shows particles with both the m-WO_3 and h_1-WO_3 structures, but also nanoparticles of a new phase, also hexagonal, but with a much larger unit cell that that of h_1-WO_3 (Tilley and Jefferson, 1999). A micrograph of a particle of this phase is shown in *Figure 5.7(c)*, and a schematic diagram of the structure, which has been confirmed from image simulation experiments, in *Figure 5.7(d)*. This new phase, which has been designated h_2-WO_3, may be regarded as intermediate between the known monoclinic and hexagonal forms, in that it contains the hexagonal tunnels of the latter separated by groups of four octahedra from the former. A similar configuration has been observed in bulk specimens of $Sb_{0.2}WO_3$, although in the latter the separation of the hexagonal tunnels by elements of the m-WO_3 structure is only in one direction (Dobson *et al.*, 1987). It is believed that the tunnels of the h_1-WO_3 structure form around H_3O^+ ions which are present at low pH:raising the pH effectively reduces their concentration and ensures that the monoclinic structure begins to form. At intermediate pH values, however, the hexagonal tunnels will still form but their density is reduced, and the space between them is filled with elements of the monoclinic structure. A whole series of intermediates is therefore possible, but although disordered nanoparticles have been noted so far only the h_2-structure has been observed in a perfect arrangement, and that only in relatively small nanoparticles. The element of the h_2-structure derived from m-WO_3 is heavily distorted and therefore extremely strained, and it seems highly probable that in bulk material, this strain would be intolerable, although it can be taken up by surface relaxation in a nanoparticle. In the original solution, these nanoparticles almost certainly constitute large polyanions. It is therefore quite possible that other hitherto unknown structural variants can be created using this technique.

5.4 Conclusions

Studies carried out in the last few years show conclusively that, especially in metal oxides, the importance of surface energy can be such that particle structures which are not the same as those of bulk materials will result. If this is the case, we cannot regard these nanoparticles as small crystals of bulk material, and the physical and chemical properties of the latter can no longer be assumed to apply. This is very well illustrated by the enhanced solubility of La_2O_3 in nanoparticulate CeO_2, and even more so by the surface metallization of the spinel-based oxides such as γ-Al_2O_3 and Fe_3O_4, and the manner in which platinum, tin and rhenium appear to dissolve in the surface layer of the former. In view of phenomena like this, we must therefore look anew at the chemical reactivity of all nanoparticles, which effectively constitute a new, and hitherto unexplored, area of solid-state chemistry.

One preconceived idea which must be challenged is the persistent belief that nanoparticles are basically crystalline in nature. By definition, a crystal consists of a unit cell or motif repeated on a lattice. All the available evidence points to the fact that in most nanoparticles the structure is not periodic, but varies continually from the centre to the surface of the particle. Consequently nanoparticles are probably better regarded as constituting large inorganic macromolecules, with molecular, rather than crystalline properties. The determination of the exact nature of these properties, which of necessity must depend on the particle size, is a daunting task, and one which must wait until effective methods of separation of the different particle sizes have been developed. There is a very obvious analogy here with the study of organic macromolecules. In the early days these were characterised purely by size and molecular weight, and it was only following the development of efficient methods of separation that their detailed structures and properties were then elucidated. In view of their potential, and so far unknown impact on the environment, both physical *and* biological, similar progress in the field of inorganic nanoparticles might seem to be advisable.

References

Bevan, D.J.M. (1955) Ordered intermediate phases in the system CeO_2-Ce_2O_3 *J. Inor. Nuc. Chem.* **1**: 49–59

Bevan, D.J.M. (1973) '*Comprehensive Inorganic Chemistry.*' Pergamon Press, Oxford p. 522.

Brauer, G. (1964) '*Progress in The Science and Technology of the Rare Earths.*' Pergamon Press, New York. 1.

Brinker, C.J. and Scherer, G.W. (1990) '*Sol-Gel Science: The Physics and Chemistry of Sol-gel Processing.*' Academic Press, New York.

Cowley, J.M. and Moodie, A.F. (1957) The scattering of electrons by atoms and crystals I A new theoretical approach. *Acta Cryst.* **10**: 609–619.

Dobson, M.M., Hutchison, J.L., Tilley, R.J.D. and Watts, K.A. (1987) The structures of intergrowth tungsten bronzes of Ba, Sn, Pb and Sb. *J. Solid State Chem.* **71**: 47–60.

Ekstrom, T. and Tilley, R.J.D. (1980) The crystal chemistry of the ternary tungsten oxides. *Chemica Scripta* **26**: 535–546.

Figlarz, M. (1989) New oxides in the WO_3-MoO_3 system. *Progress in Solid State Chem.* **19**: 1–46.

Frank, J. (1973) The envelope of electron microscope transfer functions for partially coherent illumination. *Optik*, **38**: 519.

Gerand, B., Nowogrocki, G., Guenot, J. and Figlarz, M. (1979) *J. Solid State Chem.* **29**: 429–434.

Goodman, P. and Moodie, A.F. (1974) Numerical evaluation of N-beam wave functions in electron scattering by the multislice method. *Acta Cryst.* **A30**: 280–290.

Gribelyuk, M.A., Harris, P.J.F. and Hutchison, J.L. (1994) Plan-view and profile imaging of sulfided platinum particles. *Phil. Mag.* **69**: 655–669.

Grinton, G.R. and Cowley, J.M. (1971) Phase and amplitude contrat in electron micrographs of biological materials. *Optik* **34**: 221.

Gschnieder, K.A. and Eyring, L. (1979) *Handbook on the Physics and Chemistry of Rare Earths*, Vol 3. North Holland, Amsterdam.

Hanzen, K.J. (1971) The optical transfer theory of the electron microscope: fundamental principles and applications. *Adv. Opt. Elec. Microsc.* **4**: 1.

Harris, P.J.F. (1986) The sulfur-induced facetting of platinum catalysts particles. *Nature* **323**: 792–794.

Harris, P.J.F., Boyes, E.D. and Cairns, J.A. (1983) The sintering of an alumina-supported platinum catalysts studied by transmission electron microscopy. *J. Catalysis* **82**: 127–146.

Heinemann, K., Yacaman, M.J., Yang, C.Y. and Poppa, H. (1979) *J. Cryst. Growth* **47**: 177.

Howie, A. and Marks, L.D. (1984) Elastic strains and the energy balance for multiply twinned particles. *Phil. Mag.* **A49**: 95.

Jefferson, D.A. and Harris, P.J.F. (1988) Direct imaging of an adsorbed layer by high resolution electron microscopy. *Nature* **332**: 617–620.

Jefferson, D.A. and Kirkland, A.I. (1988) Image simulations of quasi-periodic small metal particles. *IOP Short Meetings Series* **11**: 71–88.

Jefferson, D.A., Kirkland, A.I., Reller, A., Tang, D., Williams, T.B. and Zhou, W. (1992) Bulk and surface structure of oxide particles using HREM. *Electron Microscopy 92* **2**: 611–614.

Kirkland, A.I., Jefferson, D.A. and Tang, D. (1991) A study of colloidadl oxides by high resolution electron microscopy. *Inst. Phys. Conf. Series* **119**: 39–42.

Kirkland, A.I., Jefferson, D.A., Duff, D.G., Edwards, P.P., Gameson, I., Johnson, B.F.G. and Smith, D.J. (1993) Structural studies of trigonal lamellar particles of gold and silver. *Proc. Roy. Soc. London* **440**: 589–609.

Kirkland, A.I., Jefferson, D.A., Tang, D. and Edwards, P.P. (1991) High-Resolution Image Simulations of Small Metal Particles. *Proc. Roy. Soc. London* **A434**: 279–296.

Marks, L.D. (1983) Modified Wulff constructions for twinned particles. *J. Cryst. Growth* **61**: 556.

Marks, L.D. and Smith, D.J. (1981) High resolution studies of small particles of gold and silver: 2 Single crystals, lamellar crystals and poly-particles. *J. Cryst. Growth* **54**: 425.

Marks, L.D. and Smith, D.J. (1983) HREM and STEM of defects in multiply twinned particles. *J. Microscopy* **130**: 249–261.

McHale, J.M., Auroux, A., Perotta, A.J. and Navrotsky, A. (1997) Surface energies and thermodynamic phase stability in nanocrystalline aluminas. *Science* **277**: 788–791.

Morris, B.C., Flavell, W.R., Mackrodt, W.C. and Morris, M.A. (1993) Lattice-parameter changes in the mixed-oxide system cel-xlaxo2-x/2 – A combined experimental and theoretical-study. *J. Mater. Chem.* **3**: 1007–1013.

Narain, K. (1995) High Resolution Transmission Electron Microscopy studies of the sulphiding of palladium particles. *PhD Thesis*, University of Cambridge.

Perrichon, V., Laachir, A., Bergeret, G., Fréty, R., Tournayan, L. and Touret, O. (1994) Reduction of cerias with different textures by hydrogen and their reoxidation by oxygen. *J. Chem. Soc. Faraday Trans.* **90**: 773–781.

Reed, T.B. (1970) *The Chemistry of Extended Defects in Non-Metallic Solids.* North Holland, Amsterdam, pp. 21–36.

Scherzer, O. (1949) The theoretical resolution limit of the electron microscope. *J. Appl. Phys.* **20**: 20.

Sinha, K.P. and Sinha, A.P.B. (1958) *Z. Anorg. Chem.* **293**: 228–232.

Smith, D.J., Bursill, L.A. and Jefferson, D.A (1986) Atomic imaging of oxide surfaces. *Surf. Sci.* **175**, 673–683.

Smith, D.J. and Marks, L.D. (1981) *Phil. Mag.* **A44: 735.**

Stokes, A.R. (1948) *Proc. Phys. Soc. London* **61**: 382–395.

Tilley, E.E.M (1997) Synthesis and characterisation of nanocrystalline metal oxides. *Ph.D Thesis*, University of Cambridge.

Tilley, E.E.M. and Jefferson, D.A. (1999) Nanoparticles of a new form of tungsten trioxide. *J. Mat. Chem.* (in press).

Uppenbrink, J., Kirkland, A.I., Tang, D. and Jefferson, D.A. (1990) High resolution profile imaging studies of sulphur overlayers on platinum (100) surfaces. *Surf. Sci.* **238**: 132–141.

Uppenbrink, J., Kirkland, A.I., Wales, D., Jefferson, D.A. and Urban, J. (1992) Stationary points of model symmetric and asymmetric decahedra. *Phil. Mag.* **B65**: 1079–1096.

Wells, A.F. (1984) *Structural Inorganic Chemistry* (4th edn.). Oxford University Press, pp. 516–573.

6 Lung deposition of diesel particulates

John McAughey

6.1 Introduction

6.1.1 Health effects

There has been increasing concern in recent years, with respect to acute health effects and mortality from episodes of particulate pollution. A number of reviews have been conducted which indicate that the more significant health effects may include exacerbation of diseases of the cardio-respiratory system and asthma (e.g. reviewed by COMEAP, 1995, 1998). Particle exposures typically also include exposure to gaseous co-pollutants such as NO_x, SO_x and CO.

It is important to establish a plausible biological mechanism linking particle exposure to health effects, both in terms of establishing pre-clinical markers for susceptible sub-populations, and for informing policy on reducing exposure and hence risk. As part of this exercise, it is important that exposure be broken down to establish which exposure fractions of the ambient aerosol affect health. By better defining exposure, particle dose to discrete regions of the lung may be better established, as a first step to establishing potential for systemic uptake, or for the manifestation of disease at these specific sites within the lung.

Assessing the degree of exposure to environmental particulates must take into account individual or population dose of the appropriate fraction of the ambient aerosol. Dose relates to the deposition, uptake and retention of each component of the ambient aerosol, which may be affected by a number of factors. These include: (1) particle size, shape and solubility (the single most important factor); (2) chemical composition of the aerosol; (3) local environment, i.e. temperature, humidity, ventilation and lifestyle; (4) anatomical and physiological factors such as inter-

Particulate Matter, edited by R.L. Maynard and C.V. Howard

subject variability, gender, age (physique) and disease; (5) personal exposure relative to area sampling concentrations.

Interaction of these variables gives rise to a complex matrix in terms of absolute dose determination at the population level but a pragmatic approach may be adopted to estimate how these factors may influence relative dose, and hence the sensitivity of dosimetric models to these variables.

6.1.2 Size distribution of ambient particles

In practice, particles within the PM_{10} fraction can be classified into three principal modes, related to the physical processes controlling their formation, each mode having a relatively distinct mechanism of deposition within the lung.

- A nucleation mode (0.01–0.1 μm diameter, QUARG, 1995) principally emitted from processes involving condensation of hot vapours, including vehicle emissions. Particles within this size range are most likely to deposit in the lung by Brownian motion. Diesel particulate is typically found in this mode and size range.
- An accumulation mode (0.1–1.0 μm diameter) comprising mainly particles which have grown from the nucleation mode arising mainly from anthropogenic sources. It has been suggested that the accumulation fraction may itself be bi-modal with a condensation mode (~0.2 μm diameter) and a droplet mode (~0.7 μm diameter) (QUARG, 1995). This mode is most likely to be resuspended by human activity with a low subsequent settling velocity, hence the longest lifetimes in the atmosphere. Deposition in the lung in this size mode occurs by a combination of impaction and sedimentation but with low deposition efficiency. However, particles in this mode will penetrate more deeply into the lung.
- A coarse mode (>1 μm diameter) produced principally by mechanical attrition covering soil dusts, industrial dusts and the like. Lung deposition for this mode is principally by impaction. It should be noted that this mode dominates PM_{10} measurements (by mass) such that variations in the other particle modes may be swamped.

6.1.3 Source apportionment of ambient particles

The formation route of many aerosols from specific sources may determine the size range where particular components appear. Examples reported include:

- diesel particulate emissions (primary nucleation mode) of 0.04–0.05 μm;
- polyaromatic hydrocarbons (PAHs) associated with a 0.075–0.12 μm band;
- indoor combustion aerosols associated with a 0.3–0.4 μm range;
- natural gas and propane flames associated with a 0.01–0.1 μm range;
- environmental tobacco smoke 0.1–0.7 μm range.

Chemical composition of the aerosol, rather than mass concentration alone, must ultimately be considered when fully assessing dose to the population. Data on aerosol composition with respect to size fractions in the Los Angeles area have been reported for anions and organic species. However, a full consideration of composition aspects is beyond the scope of this chapter.

6.1.4 Particle number vs. particle mass distribution

PM_{10} is frequently defined as particulate matter of less than 10 μm aerodynamic diameter, but more accurately refers to a sampling convention of a size selective inlet with a 50% efficiency cut-off at 10 μm aerodynamic diameter, and is based on a sampling definition of thoracic fraction (International Standards Organisation, 1994).

Aerodynamic particle sizes are a convention used to allow direct comparison of aerosols of differing density, with respect to their likely deposition behaviour in the respiratory tract. A density correction is made to compare particles versus spheres of unit density which will sediment at the same rate as the real particle. Thus, large particles of low density will have an aerodynamic diameter less than the real diameter and dense aerosols an aerodynamic diameter greater than the real diameter. This convention is particularly useful for particles greater than 0.1 μm, where sedimentation and impaction are the principal mechanisms of deposition.

However, particle size distributions can be determined by number, equivalent surface area or mass – where particles in the nucleation mode are considered, a number distribution may be more appropriate. Alternative distributions may be more appropriate in determining mechanisms where dose is determined by number (e.g. radiation, fibre toxicology), or by delivery of adsorbed material from available surface area (e.g. nicotine in cigarette smoke), rather than particulate mass, which will be dominated by the coarse mode (by mass, 1×10 μm particle $= 1000 \times 1$ μm particles $= 10^6 \times 0.1$ μm particles $= 10^9 \times 0.01$ μm particles).

6.1.5 Vehicle emissions by particle size

The particle size distribution for vehicle emissions from a series of gasoline-, diesel- and gas-powered vehicles has been measured. Samples were collected from a Constant Volume Sampling (CVS) system on a chassis dynamometer and measured using a Scanning Mobility Particle Sizer (SMPS; TSI Inc., USA). Examples of distributions are shown in Figure 6.1 for a light-duty diesel vehicle. It is clear that the particles formed are typically 0.1 μm (100 nm) and below, and will contribute to the nucleation mode in the ambient distribution. It is important to note that particle flux may be very high (10^{12}–10^{14} particles/cc over an emissions drive cycle).

A subsequent study in Oxford city centre involved the measurement of particle number versus particle mass (by TEOM with a PM_{10} inlet). This

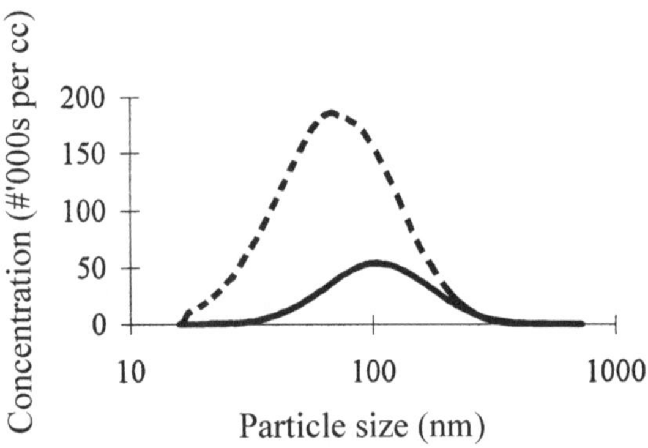

Figure 6.1. Vehicle emissions – number size distribution,
- - - CVS 70 kph,
—— MiniCVS 70 kph.

confirmed that significant particle number levels ($>10^6$ per cc) can occur in urban centres (APEG, 1999).

6.2 Lung dose of diesel particulate

6.2.1 Basis of human lung deposition models

Human lung deposition models are based typically on estimation of the physical behaviour of aerosol particles in a simplified anatomic model of the lung, that is as a series of branching cylinders with bifurcations of known angles allowing calculation of deposition by impaction, sedimentation and diffusion. These models are generally weighted to take account of experimental data. The most widely used models are based on a simplified branching model of the respiratory tract as published by Weibel (1973), the most widely used models being those published by the International Task Group on Lung Dynamics, and more recently, ICRP 66 (1994).

The ICRP 66 model has been validated by particle deposition measurements in humans under controlled exposures, typically using radio-labelled particles to assess regional deposition in the various lung compartments. In general, the data are in good agreement with model predictions, although a degree of inter-subject variability is observed. The ICRP 66 model extends the scope of earlier models and considers variation in deposition induced by factors such as anatomy/physiology (including age, gender and race), ventilation rate and pattern, nature of exposure and particle size.

6.2.2 Anatomy and physiology of the respiratory tract

The respiratory tract in the lung model is greatly simplified with division into four main regions: extrathoracic (ET), bronchial (BB), bronchiolar (bb) and alveolar-interstitial (AI) with simplifying assumptions for clearance broken down into simple considerations of mechanical clearance and translocation. Symmetrical airway branching and ventilation are assumed throughout the deposition models, although this is not observed in real life.

- ET regions: The choice of nasal vs. oral breathing is largely determined by the relative resistance of the two pathways with respect to ventilatory needs. Nasal breathing, which conditions the temperature and humidity of the inhaled air and filters larger particles, predominates in healthy subjects at rest, with oronasal breathing increasing in response to exercise. The anatomical structure of this region is characterized by airflow resistance of approximately 0.15 kPa $l^{-1}s$, equivalent to half the total airway resistance. In most healthy individuals, a nasal cycle is observed, with an anatomical and resistance asymmetry through nasal congestion and decongestion. This causes an approximate 3 h cycle with airflow distribution changing as much as 20–80% between nostrils without subjective discomfort, the effect disappearing in elderly people and during exercise. The introduction of turbulent flow through this asymmetry is likely to increase deposition efficiency. Studies of deposition efficiencies in the ET region using a combination of volunteer studies and studies involving the use of nasal casts have been extensively reviewed by Cheng *et al.* (1991). However, there remains a need for further comparative data for nasal and oral breathing.
- BB and bb regions: These regions comprise what is known as the classical 'tracheobronchial tree' or 'conducting airways' where inspired gases are transported but do not interchange with blood. The model is based on a branching structure from Generation 0 (trachea) to Generation 26 (last of branches), although the number of generations present varies significantly in different parts of the lungs. The bronchial region comprises airways greater than 1 mm diameter (to about Generations 8–10) with bronchiolar airways having diameters of less than 1 mm. The region is characterized by a rapid clearance phase for deposited particles via a mucociliary escalator, which is generally complete within 24 h. A slower compartment of bronchial clearance has been observed in both animals (Patrick, 1983) and man (Stahlhofen, 1989).
- AI: This comprises the gas-exchange region comprising respiratory bronchioles through to alveolar sacs. Alveolar deposition in this region is modelled by assuming an inflatable 3/4 spheroid as a model of the alveolus with deposition by a combination of settling velocity and diffusive flux. Clearance of particles from this region is principally by phagocytosis with substantially longer clearance times, dependent to some extent on particle deposition. In a review by Bailey *et al.* (1991), slow half-times of clearance range from 300 days to a number of years, which has implications for long-term particle retention.

The principal criteria of the ICRP lung model have been adopted in developing LUDEP (Birchall *et al.*, 1991), a microcomputer-based model used to predict regional particle deposition in the human lung, offering a useful means of estimating differences in deposition pattern across different population groups and under different conditions.

6.2.3 Mechanisms of particle deposition

Five principal mechanisms of particle deposition exist based on the physical properties of the aerosol, as described below.

- Impaction, due to the inertia of the particles, occurs when the air undergoes a change of direction such as in the pharynx or at a bifurcation of the airways. This is important for the deposition of larger particles (> 1 μm) in the upper airways and the nasopharyngeal region. Clearance of such particles, where insoluble, occurs primarily via the mucociliary escalator. There is additional experimental evidence from deposition measurements in early generations of human lungs that vortices are set up downstream from birfucations leading to regions of concentrated particle deposition rather than uniform deposition for that particular airway generation (Martonen *et al.*, 1992). Deposition by impaction increases as particle momentum increases, that is, with increased air flow and particle size. Bronchial dimensions are strongly dependent on physique; on average the side to side diameter of the adult male trachea is about 2 cm, that of the adult female about 1.5 cm. Since the flow of air, and hence the momentum of any particle, is inversely proportional to the square of the airway diameter, deposition by impaction in the upper airways will show clear differences between men and women as impaction efficiency increases at a given flow and this has been observed (Pritchard *et al.*, 1986).
- Sedimentation is an important mechanism of deposition in parts of the respiratory tract where air flow velocity is low: it plays an important part in deposition in the bronchioles. The efficiency of sedimentation as a mechanism of deposition is dependent upon the terminal settling velocity of the particle concerned which is dependent on the square of the particle radius. In practice, deposition efficiency is low relative to impaction, although it becomes more significant for particle sizes in the 0.5–1 μm region. The low efficiency of deposition by this mechanism is reflected by the minimum of particle deposition, in adults, occurring for 0.5 μm particles.
- Brownian motion: Particle deposition by Brownian motion occurs due to the bombardment of small particles (< 0.1 μm) by air molecules. This is most important for the deposition of small particles which penetrate to the alveolar region. This region has a large surface area allowing rapid dissolution and uptake of soluble particles. For insoluble particles, macrophage clearance predominates. The rate of subsequent clearance of lipid soluble compounds tends to be related to

their molecular weight, with more rapid clearance at molecular weights of less than 300 Da (Henderson *et al.*, 1988).

- Interception is a term used to describe deposition of particles as a result of their coming into contact with airway walls where the direction of movement of the particles does not deviate as rapidly as that of the air stream. Interception is an important mechanism for the deposition of fibres, particularly at airway bifurcations.
- Electrostatic deposition: Charged particles in an aerosol will repel each other leading to an increase in migration towards airway walls, to which they may also be attracted as a result of charge induced on these walls. However, in practice, most ambient particles become neutralized by ions in air and in real terms, electrostatic deposition plays a small role in determining deposition in the respiratory tract.

For non-ideal particle behaviour, shape and solubility of the inhaled particle may also have an important role.

- Particle shape: The majority of environmental aerosols encountered by the population will be sufficiently spherical to be considered to behave as spherical particles in deposition models. The exceptions are fibrous aerosols, such as asbestos and man-made mineral fibres (MMMFs). Fibrous materials tend to align with airflow so that aerodynamic diameter is not strongly dependent on length, and fibres of up to 360 μm in length have been found to penetrate to the human alveoli. It is not possible for lung clearance mechanisms to operate on fibres of this type, and defence mechanisms rely on depositing a proteinaceous covering on the fibres.
- Particle solubility: Solubility may be significant for some components of the ambient aerosol where, on entering the high humidity of the respiratory tract, water vapour will condense on particles and form droplets of solution. If soluble, the particle hydrates, eventually forming a solution. As the vapour pressure of a solution is less than that of the pure solvent, the particle may swell to a droplet several times its initial size, a phenomenon known as hygroscopic growth. In the most extreme case for sodium chloride, an aerodynamic growth factor of 4.5 is observed leading to a marked shift in the deposition pattern of sub-micron particles; lower growth factors are reported for sulphate based aerosols. Owing to the maritime climate, the majority of the UK ambient aerosol is hygroscopic. Soluble particles which deposit will dissolve in the lining fluids of the respiratory tract and so will not be cleared by the mechanisms involved in the clearance of insoluble particles.

6.2.4 Efficiency of diesel particle deposition – assumptions for calculations

Sample calculations have been conducted using data from the APEG (1999) report, which suggests that, on average, one-third of the background annual average PM_{10} concentration is attributable to primary combustion sources.

These data were input into a series of examples of deposition as calculated by the LUDEP computer program of the ICRP 66 lung model for daily exposures. Values chosen were:

PM_{10}:	Annual mean	27 μg m^{-3}
	Attributable to primary combustion emissions (approx)	9 μg m^{-3}
Subjects:	Adult male	0.78 m^3 h^{-1} mean ventilation rate
	1-year-old child	0.19 m^3 h^{-1} mean ventilation rate
Particle composition		70% 0.2 μm MMAD
By mass for diesel particles		30% 2.0 μm (coarse mode)

The size composition choice is somewhat arbitrary, but is based on operational experience of measurement over a variety of fuel and engine technologies.

6.3 Dose calculations

6.3.1 Daily mass deposition

Mean daily exposures for an adult, therefore, are:

$$\text{Concentration} \times \text{volume inhaled h}^{-1} \times \text{Time}$$
$$9\ \mu\text{g m}^{-3} \times 0.78\ \text{m}^{-3}\ \text{h} \times 24\ \text{h} = 170\ \mu\text{g}$$

For a 1-year-old child, mean daily exposures are:

$$27\ \mu\text{g m}^{-3} \times 0.19\ \text{m}^3\ \text{h}^{-1} \times 24\ \text{h} = 40\ \mu\text{g}$$

Using LUDEP deposition fractions, calculated results for total and regional deposition are shown in *Table 6.1.* Daily particle retention in the adult averages 68 μg, approximately 40% of which penetrates into the thoracic region of the lung. For the 1-year-old child, retention averages 21 μg, such that total deposition efficiency is greater than in the adult, but penetration to the thoracic region is reduced.

6.3.2 Daily particle deposition (by number)

For particle deposition by number, it is assumed that all particles are 0.2 μm aerodynamic diameter (number will be dominated by the small

Table 6.1. Daily particulate deposition (by mass) by calculation

	Extra-thoracic (μg)	Bronchial (μg)	Alveolar (μg)	Total (μg)	Deposition efficiency (%)
Adult (mean retention)	32.4	8.5	26.8	67.7	40
Child (1-year-old; mean retention)	11.9	2.6	6.1	78	52

mode) and a particle concentration of 10^5 particles cm^{-3} has been chosen as a typical urban exposure (APEG, 1999). Thus, adult daily intake (calculated for the lung can be calculated as follows:

$$\text{No of particles} = \text{concentration} \times \text{volume inhaled} \times \text{deposition efficiency (AI)}$$
$$= 100\,000\ \text{cm}^{-3} \times (780\,000\ \text{cm}^3\ \text{h}^{-1} \times 24\ \text{h} \times 0.29)$$
$$= 5.4 \times 10^{11}\ \text{day}^{-1}$$
$$(3.35 \times 10^{11}\ \text{penetrating to the alveoli})$$

One further area of interest is whether particle overloading may occur in the alveolar macrophages and estimates of daily penetration can be made. However, this assumes that all alveoli are ventilated equally. Thus, if alveolar number in the adult = 296×10^6 (ICRP 66), then number of particles deposited per alveolus = 1130 day^{-1}.

Crapo *et al.* (1982) estimate 2.16×10^{10} macrophages in the human alveolar region, implying 73 macrophages per alveolus. Uptake of particles by macrophages on this scale is not likely to lead to 'overloading'. Morrow (1994) has estimated that macrophage overloading begins when the total volume of particles ingested by a macrophage reaches 60 μm^3. The volume of a macrophage has been estimated as 1000 μm^3. Overload of a macrophage is said to be complete, i.e. mobility of the macrophage is reduced to zero, when the total volume of the ingested particles reaches 600 μm^3, i.e. 60% of macrophage volume. Öberdorster *et al.* (1992) have demonstrated macrophage overloading on ingestion by macrophages of single 10.3 μm diameter polystyrene spheres. The volume of one such sphere is 572.2 μm^3. In contrast, the volume of a particle of diameter 0.05 μm (physical size of primary diesel particle) is 0.000065 μm^3. A macrophage would therefore need to ingest 917 431 such particles before beginning to be overloaded.

6.3.3 Retention

Table 6.1 shows that approximately 27 μg diesel particulate per day penetrates to the AI region in the adult lung. Material penetrating to the alveolar region generally demonstrates multi-phase clearance which will be size- and composition dependent to some extent. If clearance is simplified assuming first order clearance kinetics, and the worst case situation is chosen (all clearance is in the slower phases) estimates of equilibrium lung burden can be calculated. A half-time of 1 year has been chosen as realistic for the carbon-rich 0.2 μm particle mode: it is recognized, however, that soluble components will be rapidly removed.

Thus, at equilibrium concentration (A_0), particle retention = particle clearance (27 μg day^{-1})

For first order kinetics:

$$A/A_0 = e^{-kt}$$
$$\ln (A/A_o) = -kt$$

where $A/A_o = 0.5$ (half-time of clearance) and $t = 365$ days is assumed, $k = 0.0019\ day^{-1}$.

Then, for $t = 1$ (i.e. clearance in 1 day):

$$A/A_0 = 0.9981 \text{ (where } A = A_0 - 27)$$
$$(A_0 - 27)/A_0 = 0.9981$$
$$A_0 = 14\,210 \ \mu\text{g}.$$

Thus, assuming a half-time clearance of 1 year, the lung burden of diesel particles at equilibrium will be approximately 14 mg; for a half-time of 36 days, the equilibrium burden would be 1.4 mg.

6.4 Conclusions

In conclusion, dose calculations have been presented in which estimates of particle deposition and retention for mass and number have been made. Environmental and vehicle emissions data suggest that consideration of diesel particle dose on a number basis is valid, due to the size of particles observed and the efficiency of their penetration to the lung.

It is noted that these calculations are underwritten by a broad base of assumptions; it is expected that these dose estimates can be better modified and validated as additional experimental data become available. Most importantly, a wide degree of biological variation is anticipated for population dosimetry. In particular, ventilation asymmetry may become increasingly significant in those with existing respiratory diseases.

References

APEG (1999) *Source Apportionment of Airborne Particulate Matter in the United Kingdom.*

Bailey, M.R., Birchall, A., Cuddihy, R.G., James, A.C. and Roy, M. (1991) Respiratory tract clearance model for dosimetry and bioassay of inhaled radionuclides. *Rad. Prot. Dosim.* **38:** 153–158.

Birchall, A., Bailey, M.R. and James, A.C. (1991) LUDEP: A lung dose evaluation programme. *Rad. Prot. Dosim.* **38:** 167–174.

Cheng, Y.S., Yeh, H.C. and Swift, D.L. (1991) Aerosol deposition in human nasal airway for particles 1 nm to 20 μm: a model study. *Rad. Prot. Dosim.* **38:** 41–47.

Crapo, J.D., Barry, B.E., Gehr, P., Bachofen, M. and Weibel, E.R. (1982). Cell characteristics of the normal human lung. *Am. Rev. Resp. Dis.* **125:** 740–745.

COMEAP (1995) *Committee on Medical Effects of Air Pollutants. Non-biological Particles and Health.* HMSO, London.

COMEAP (1998) *Committee on Medical Effects of Air Pollutants. Quantification of the effects of Air Pollution on Health in the United Kingdom.* HMSO, London.

Henderson, R.F., Bechtold, W.E., Medinsky, M.A., Fischer, J.P. and Lee, T.T. (1988) The effect of molecular weight/lipophilicity on clearance of organic compounds from lungs. *Toxicol. Appl. Pharmacol.*, **95:** 515–521.

ICRP (1994). *Human Respiratory Tract Model for Radiological Protection.* ICRP Publication 66. *Ann. ICRP* **24:** (1–3).

International Standards Organisation (1994) *Air Quality – Particle Size Fraction Definitions for Health-Related Sampling*, IS 7708. ISO, Geneva.

Martonen, T.B., Zhang, Z. and Yang, Y. (1992) Extrapolation modelling of aerosol deposition in human and laboratory rat lungs. *Inhalation Toxicol.* **4:** 303–324.

Morrow, P.E. (1994) Mechanisms and significance of 'particle overload'. In: Mohr, U. (ed), *Toxic and Carcinogenic Effects of Solid Particles in the Respiratory Tract.* Hemisphere Publishing, Washington DC, pp. 17–25.

Öberdorster, G., Ferin, J. and Morrow, P.E. (1992). Volumetric loading of alveolar macrophages (AM): a possible basis for diminished AM-mediated particle clearance. *Exp. Lung Res.* **18:** 87–104.

Patrick, G. (1983) The retention of various types and sizes of particle in the large airways of the rat: Implications for assessing the risk of lung cancer. In: Fiher, D.R. (ed) *Current Concepts in Lung Dosimetry.* US Department of Energy, pp. 66–72.

Pritchard, J.N., Jeffries, SJ. and Black, A. (1986) Sex differences in the regional deposition of inhaled particles in the 2.5–7.5 μm size range. *J. Aerosol Sci.* **17:** 385–389.

QUARG (1995) *Diesel Vehicle Emissions and Urban Air Quality.* 3rd Report. HMSO, London.

Stahlhofen, W. (1989) Human lung clearance following bolus inhalation of radioaerosols. In: *Extrapolation of Dosimetric Relationships for Inhaled Particles and Gases.* Academic Press, New York, pp. 153–166.

Weibel, E.R. (1973) Morphological basis of alveolar-capillary gas exchange. *Physiol. Rev.* **53:** 419–495.

7 The biological effects on lung epithelium of well-characterized fine particles

Roy J. Richards, Kelly A. BéruBé, Lisa Masek, Debbie Symons and Samantha A. Murphy

7.1 Introduction

Studies on workers occupationally exposed to dusts and a variety of chemicals by inhalation have shown that the mechanisms of cellular damage, in so far as they are known, follow a similar pattern. These changes will be described in this chapter with particular reference to epithelial cells lining the respiratory tract which, together with their extracellular secretions, represent the primary target for inhaled material. Respiratory tract damage in humans exposed to xenobiotics is often described after lifetime and/or persistent exposure. Similar, reproducible toxicological studies in animals can provide models to study inflammation, epithelial cell metaplasia/neoplasia or fibrosis in the respiratory tract. However, most investigations with pollutant particles involve high, sometimes overload, exposures which bear little relevance to environmental concentrations of airborne material. One of the challenges facing respiratory toxicologists at present is to define the mechanisms by which tiny masses of ultrafine (less than 100 nm diameter) or respirable (less than 2.5 μm) particles present in the atmosphere can compromise the normal functioning of the lung.

Experimental data on this aspect are presented in the second part of this chapter which examines the importance of physicochemical properties of particles to induce transient or progressive respiratory tract changes linked to alterations in the surface epithelium. Four, well-characterized preparations of particles were used in this study. The particles of major interest were diluted (mixed with ambient air) diesel

Particulate Matter, edited by R.L. Maynard and C.V. Howard

exhaust particles (DEP) because these are common ultrafine/respirable mode particles present in urban atmospheres and therefore inhaled by the majority of the population. DEP are enriched in inorganic carbon which during formation becomes contaminated with trace metal components/organic oxidation products from fuel, oil and engine parts. Man-made ultrafine/respirable carbonaceous particles, known as carbon blacks, are also commercially available. One such sample, known as M120, was chosen for investigation because like DEP it has a high proportion of ultrafine particles that aggregate. M120 however, has a simpler surface chemistry than DEP and is less damaging (by mass) to cells *in vitro*. The third dust chosen was a pure, amorphous silicon dioxide, commercially available as Cabosil. This has a high proportion of ultrafine particles, which readily aggregate like DEP but has a completely different surface chemistry (amorphous silicon dioxide) for interaction with the epithelial lining fluid and respiratory tract epithelial cells. The fourth sample, DQ12 quartz, was chosen as a positive control in that it is known to be biologically reactive and to initiate progressive lung damage. The sample is respirable but has few particles in an ultrafine (<100 nm) category and a surface of pure crystalline silicon dioxide.

A final section of this chapter is concerned with the fate and route of dispersion of DEP once deposited in small amounts by instillation at the respiratory epithelium. Some preliminary light and electron microscopical work is described which helps identify the cells that can endocytose the material and its location in the interstitium of the alveolar septum, having bypassed the epithelium. The clearance time to lymph nodes and how this tissue aggregates DEP and responds to the dust load of carbonaceous particles is also described.

7.2 Alveolar structure and cell response to toxicants

7.2.1 Functions of alveolar cells

The cellular organization of the alveolar unit as observed by electron microscopy is shown in *Figure 7.1a* and *b*. There is a rich capillary network lined by thin endothelial cells. These are divided from epithelial type I cells which line the alveolar space by a shared basement membrane. Epithelial type I cells, which line over 95% of the alveolar surface, are also extremely thin thus permitting an efficient gaseous exchange by the red blood cells in the capillary. Type I cells are thought to be of low metabolic activity and yet have been shown to contain the isoenzyme CYP2B (Dinsdale, 1998), a member of the cytochrome P450 mono-oxygenases which are important in toxicant biotransformation. Type I cells, because of their extensive surface area, are thought to be at

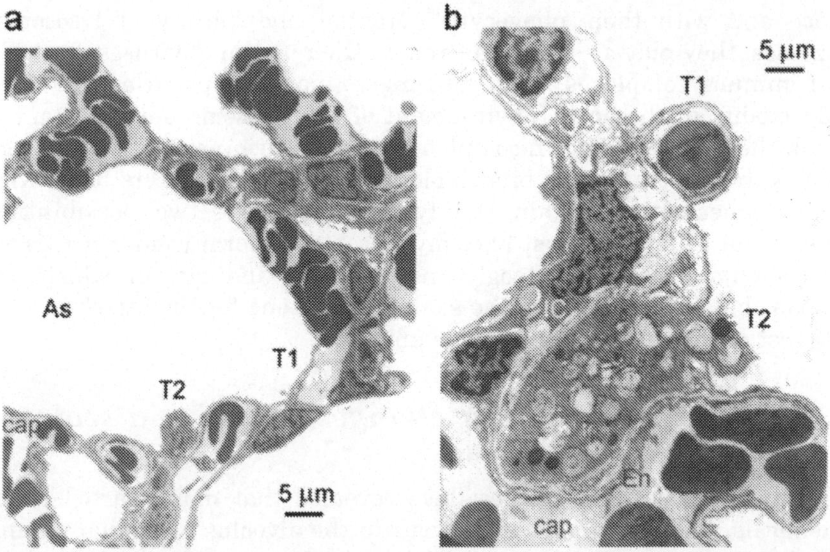

Figure 7.1. (a,b) Electron micrographs of rat normal alveolar tissue. As, alveolar space; T1, type I epithelial cell; T2, type II epithelial cell; cap, capillary containing red blood cells; En, endothelial cell; IC, interstitial cell.

risk from chemical and particle attack despite being covered by a protective epithelial lining fluid and pulmonary surfactant.

Epithelial type II cells are located in the corners of alveoli, and have an apical, microvillus-rich surface at the alveolar space which is connected by junctional complexes to type I cells. This contact apart, the cell also retains close communication via basement membranes with interstitial cells and the endothelial cells of the capillaries. The closeness of these contacts may well be related to the multiple roles exhibited by the type II cells. Firstly, they are secretory cells releasing pulmonary surfactant and/ or other products into the alveolar space. Surfactant is a lipoprotein complex which reduces surface tension and prevents alveolar collapse on expiration. The secretory components add to the protective lining fluid covering type I cells, may have a role in the immune response and are often complexed with other important antioxidant molecules such as vitamin E. In addition, type II cells synthesize and secrete components of the underlying basement membrane (Crouch *et al.*, 1987; Sage *et al.*, 1983). Secondly, type II cells are enriched with a spectrum of P450 isozymes (Devereux *et al.*, 1981; Miller *et al.*, 1989; Serabjit-Singh *et al.*, 1987). The cells also have relatively high concentrations of reduced glutathione (Horton *et al.*, 1987) and apical surface enzymes such as gamma glutamyl transpeptidase (GGT) (Dinsdale *et al.*, 1992), all of which are important in the breakdown or biotransformation of inhaled or circulating chemicals. Thirdly, some stem cells within the type II cell population have the capacity to divide when the epithelial layer is compromised and then to differentiate to type I cells, thus restoring the alveolar architecture to normality. Macrophages are found in the alveolar space and with their phagocytic potential and battery of lysosomal enzymes, they play an important role in the removal of viruses, bacteria and immune complexes. They also ingest inorganic particles that they may modify in the acidic environment of the lysosome but they cannot break these particles down completely. Alternative removal routes, such as upward movement into bronchioles or entry into the interstitium with final sequestration within the lymph nodes are two possibilities. Interstitial cells (fibroblast-like, myofibroblasts) form connective tissue components such as proteoglycans, collagen and elastin which are responsible for maintaining the skeletal backbone for the attachment of the rest of the cells in the alveolar unit.

7.2.2 Sequential cellular changes following alveolar damage

Electron microscopical studies have revealed that many particles and chemicals produce damage and repair in the alveolus by similar means, but for the purposes of illustration, most of the micrographs shown here are taken from rat lung tissue exposed to zinc chloride (Masek, 1990). It is anticipated, but not proven, that instillation of this soluble metal complex immediately compromises or bypasses the protective epithelial lining fluid and interacts with the epithelial lining cells or enters the rest of the tissue. Blebs, often full of intracellular fluid, appear on the surface of type I cells (*Figure 7.2a*) and type II cells secrete surfactant (not shown). Endothelial cells show an increase in the number of plasmalemmal vesicles (*Figure 7.2b*) (Gil, 1983), which may be linked to excessive transendothelial transport. These flask-shaped indentations of the plasma membrane (caveolae) are also found in epithelial type I cells that survive for 1 h after zinc chloride instillation (*Figure 7.2c*). Such observations equate with an increase in lung permeability and the more rapid transfer of plasma products (such as fibrinogen) and fluid to the alveolar surface where they accumulate as intra-alveolar oedema or become deposited as fibrin (*Figure 7.2d*). With mediator release from activated or dying cells this provides the stimulus for inflammation and the influx of monocytes, neutrophils or eosinophils (*Figure 7.2e*). At some focal points the tissue architecture is totally disorganized with the air spaces containing membranes of dying cells, deposited plasma products (e.g. fibrin), monocytes and other inflammatory cells (*Figure 7.2f*).

The role of replacing the epithelium is taken on by stem cells within the type II cell population (*Figure 7.3a*). In the absence of further toxicant damage such cells become flattened along the basement membrane (*Figure 7.3b*), gradually lose their surfactant-containing lamellar bodies and become type I-like cells in order to restore normal lung architecture. A critical stage in the damage/repair process occurs at this point. If the toxicant is aggressive and inflammation and epithelial cell damage persist, then the degree of re-epithelialization can become extensive and possibly take on a different form. An example of the progressive changes

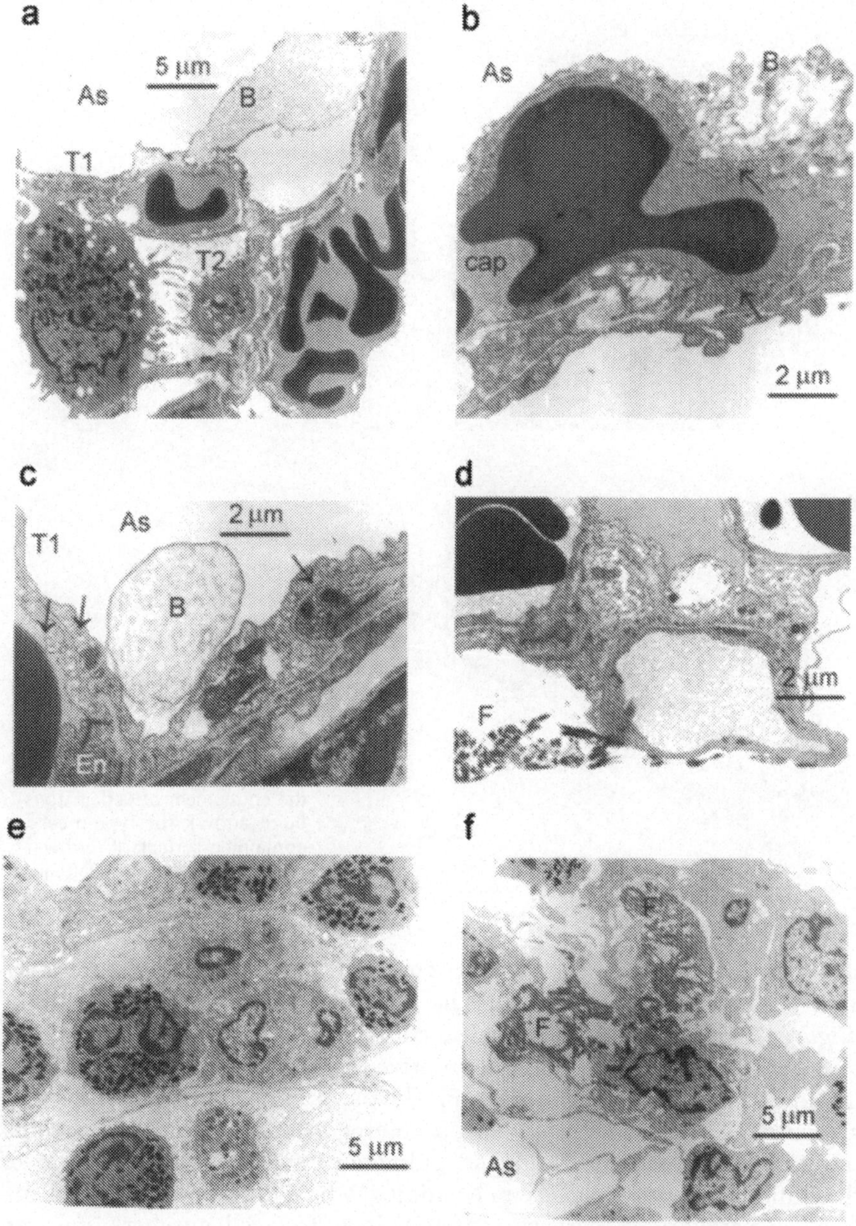

Figure 7.2. Sequential stages of alveolar damage and inflammation following instillation of zinc chloride. Key as for *Figure 7.1*. (a) Extensive blebbing (B) to epithelial type I cells. (b) Blebbing in type I cells and evidence of an increase in plasmalemmal vesicles in endothelial cells (arrows). (c) Both endothelial and type I epithelial cells show an increase in plasmalemmal vesicles (arrows). (d) Accumulation of fibrin (F) in the alveolar space. (e) Interstitial and alveolar surface accumulation of inflammatory cells. (f) Extensive damage to alveolar architecture with accumulation of cell debris, fibrin and monocytic cells.

a

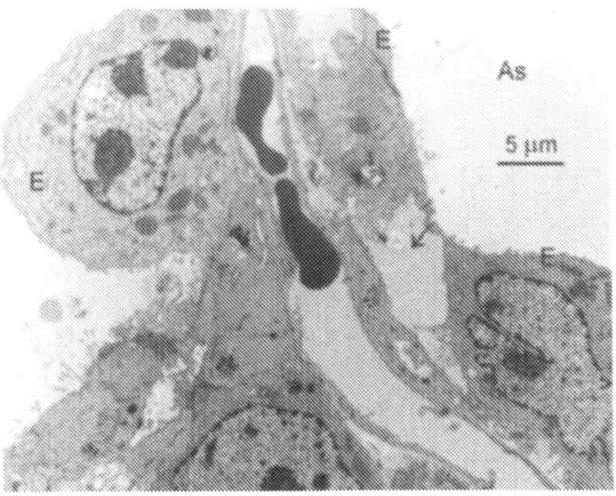

b

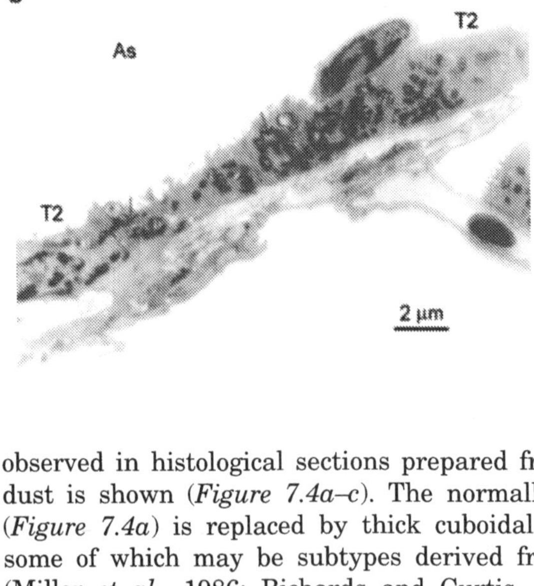

Figure 7.3. Epithelial repair. Key as for *Figure 7.1*. (a) The damaged surface epithelium is replaced by stem (type II?) cells (E) and in one area this results in the entrapment of oedematous fluid (arrow). (b) Type II cells, containing surfactant (arrows), become progressively flattened on top of a basement membrane.

observed in histological sections prepared from rats exposed to quartz dust is shown (*Figure 7.4a–c*). The normally thin (type I) epithelium (*Figure 7.4a*) is replaced by thick cuboidal cells (*Figure 7.4b* and *c*), some of which may be subtypes derived from epithelial type II cells (Miller *et al.*, 1986; Richards and Curtis, 1984). The normally clear alveolar spaces become filled with a heterogeneous mixture of free cells, cellular debris, oedema fluid and pulmonary surfactant. The extent of this inflammatory response can be adjudged by a differential count of the free cell population. Increase in macrophage free cell numbers and size heterogeneity is indicative of an inflammatory response. If the free cell population also contains a high proportion of neutrophils then the inflammation is more severe and the effects of the toxicant agent likely to be more persistent. Under extreme conditions of toxicant overload (e.g. the failure to clear high mass depositions of bioreactive particles from the alveolar surface or interstitium), alveolar proteinosis (accumulation of surfactant in the alveolar spaces) and fibrogenesis or fibrosis is seen

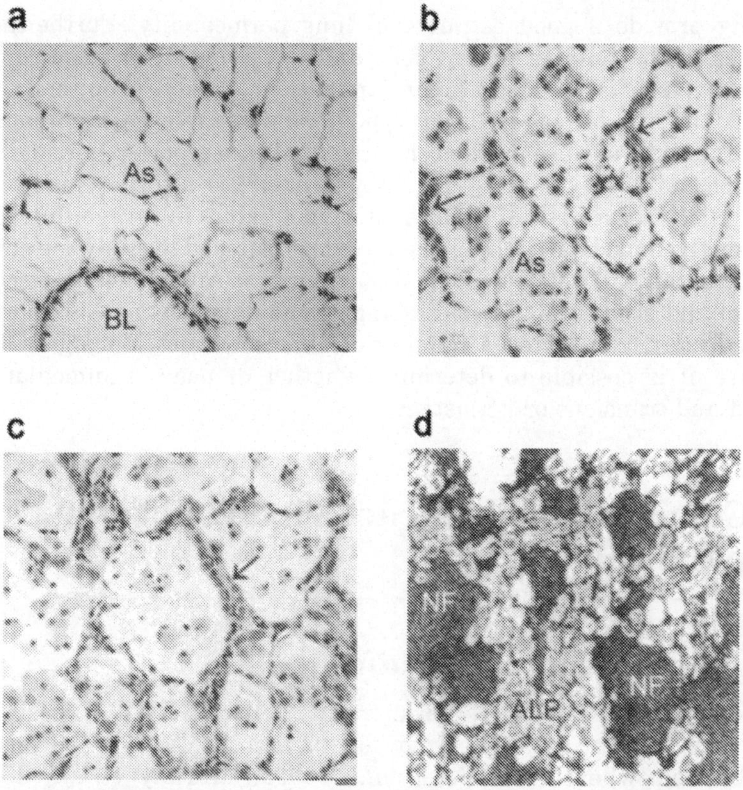

Figure 7.4. Progressive histopathological changes in rat lung exposed to quartz. (a) Normal lung architecture with thin epithelial lining cells and debris-free alveolar surfaces; As, alveolar space; BL, bronchiolar lumen. (b) Extensive alveolar surface inflammation with free cell accumulation and focal formation of cuboidal epithelium (arrows). (c) Extensive replacement of alveolar surface with thick cuboidal cells. (d) Areas of nodular fibrosis (NF) interspersed with other areas of alveolar lipoproteinosis (ALP).

(*Figure 7.4d*) (Richards and Curtis, 1984). Fibrosis may be considered irreversible and a last ditch attempt at repair and enclosure or tissue isolation of bioreactive particles (Richards *et al.*, 1991).

7.2.3 Quantitative assessment of pulmonary damage

The histopathological and electron microscope studies of lung tissue provide an excellent but subjective visual assessment of damage/repair. The application of morphometric techniques can provide semi-quantitative evaluation but can be very time consuming and assess relatively small areas of the whole lung tissue. Unless toxicants are used in overload doses there is no 'blanket' effect and lesions are often focal in distribution. A variety of other simple techniques can be applied to quantitatively assess the acute effects of toxicants, including particles, in experimental animals (Murphy *et al.*, 1998). Increases in lung wet weight/body weight ratio and protein in lavage fluid washed from the surface of the lung provide a good estimate of lung permeability. Furthermore, increases in the number of free cells and the proportion of neutrophils provide evidence for the degree of inflammation. Markers of epithelial cell integrity can also be assessed. For type II cells, the release of pulmonary surfactant and a surface marker enzyme, GGT, are reported here. Others have reported the use of $rT1_{40}$ (McElroy *et al.*, 1997), a surface marker for type I cells. Assessment of inflammatory mediators in lavage fluid and, increasingly, molecular techniques, have been applied to study the effects of particles but will not be dealt with further here. An important aspect of these animal studies is to avoid a single time point for measurement of acute effects. By studying a number of different times, during or after exposure, it is possible to determine whether damage is immediate or delayed and transient or persistent.

7.3 Comparative bioreactivity of different particles

7.3.1 Particle characterization

Four samples, M120 carbon black, freshly collected DEP, Cabosil and DQ12 quartz were characterized for physicochemical properties as described elsewhere (BéruBe *et al.*, 1999a; Murphy *et al.*, 1998). Individual particles (spherulites) have a mean diameter of 50, 30, 7 and 440 nm for M120, DEP, Cabosil and quartz, respectively. However, the individual particles of the first three dusts readily aggregate in water or the carrier solution (0.15 M NaCl) used for instillation into animals. A size distribution from image analysis of at least 1000 particle aggregates, together with their morphological form, is shown in *Figure 7.5*. Each sample has a different proportion of ultrafine (<100 nm), fine (100 nm–2.5 μm) or coarse (>2.5 μm) particles and they therefore differ in their total reactive surface area. Cabosil and DEP have similar size distributions with 74% and 79%, respectively, of particle (aggregates) below 300 nm diameter. All the samples also have a different surface chemistry, although electron probe X-ray microanalysis (EPXMA) has shown that Cabosil and quartz (which has been acid washed) are pure silicon dioxide (Murphy *et al.*, 1998). M120 is mostly carbon but has traces of sulphur and chloride detected by EPXMA. The DEP sample is also predominantly carbon but has an array of trace metal contaminents (Na, Mg, Al, Si, P, S, Ca, Fe and Zn) (BéruBé *et al.*, 1999b; Chapter 6).

7.3.2 Particle dose and administration

The particles were administered by intratracheal instillation to halothane-anaesthetized, specific pathogen-free, CD1 male rats (Charles

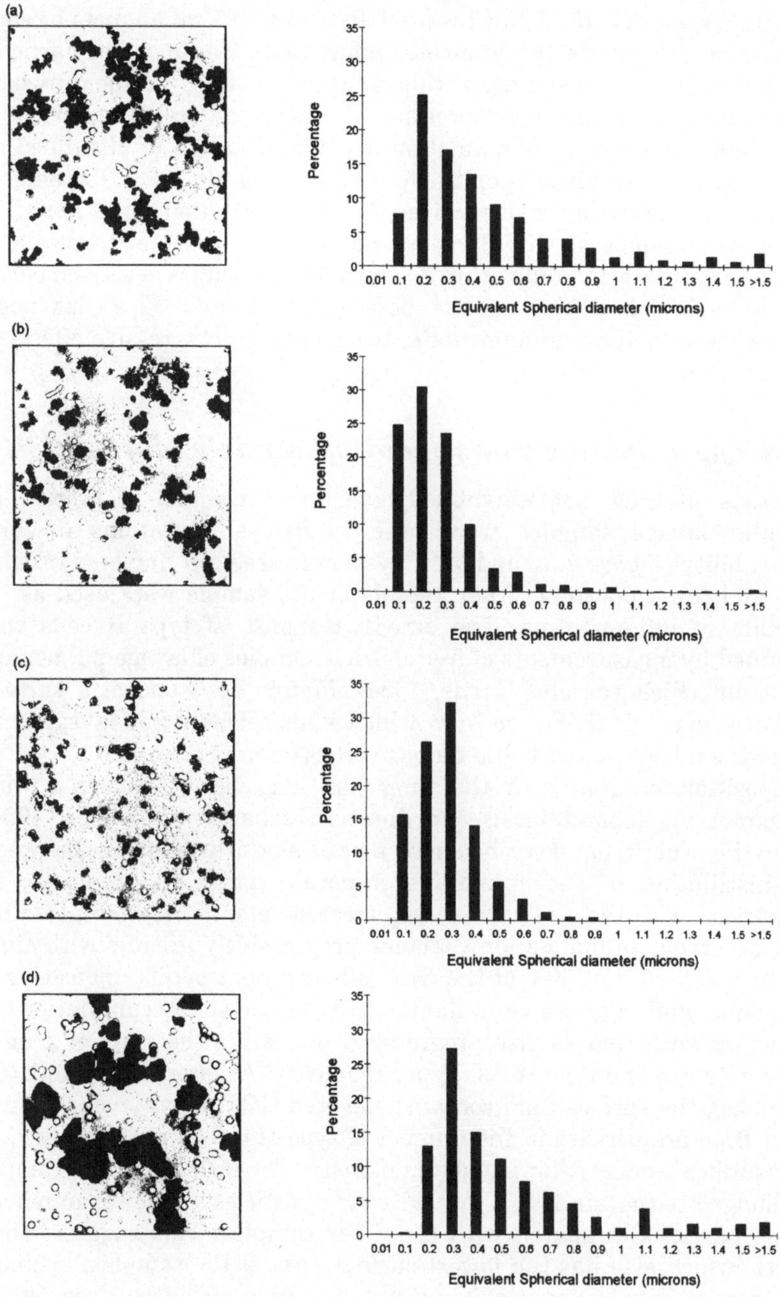

Figure 7.5. General morphology and size distribution of dust samples used in instillation studies. The (white) circular holes of the collection filter are approximately 200 nm and thus provide a guide to the diameters of particles or particle aggregates. (a) M120 carbon black; (b) diesel exhaust particulates; (c) Cabosil; (d) DQ12 quartz.

River) of approximate body weight of 200 g. Sonicated particle suspensions in 0.4 ml of 0.15 M NaCl were delivered in 0.1 ml aliquots at each inspiration just above the bronchial bifurcation. The mass of particles selected for delivery was 1 mg as this is the lung loading for the rat which is considered necessary for depression of clearance (Green and Watson, 1995) and initiation of inflammation and translocation of ultrafine/fine particles across the alveolar epithelium (Murphy *et al.*, 1998; Oberdörster *et al.*, 1992). Control animals received the carrier solution alone. For all of the above samples at this dose, animal recovery from anaesthesia is almost immediate and full within 2–3 min. In the studies described below rats were killed at time points between 48 h and 30 weeks post-instillation to determine immediate, transient and progressive effects of the particles.

7.3.3 Quantitative assessment of particle bioreactivity

Increases in lung wet weight/body weight ratios and of protein in acellular lavage samples were used to assess alterations in lung permeability (*Figure 7.6a* and *b*). Elevation in free cell numbers (*Figure 7.6c*) and the proportion of neutrophils in the sample were used as an estimate of inflammation. The activity/response of type II cells was examined by measurements of freeze-dried samples of lavage pulmonary surfactant (Richards and Curtis, 1984) (*Figure 7.6d*) and of a surface marker enzyme GGT (*Figure 7.6e*), which when released into lavage fluid provides evidence of epithelial damage (Hooftman *et al.*, 1988).

Deposition of quartz in the lung has long been associated with inflammation, lipoproteinosis and fibrosis (Richards and Curtis, 1984) and in the experiment described here, acts as a positive control. One week post-instillation of the crystalline mineral significant increases, in comparison with those of carrier only treated rats, in lung permeability were observed. Inflammation becomes progressively greater with time and between 40 and 60% of the free cells are neutrophils, indicating a persistent and aggressive inflammatory response. Accumulation of surfactant with time is also progressive, indicating the start of a mild lipoproteinosis and type II cell hyper-reactivity. At later time points (6–12 weeks), the surface fluid contains elevated GGT activity which could result from an increase in the number of type II cells and the release of their surface marker following quartz damage. The instilled quartz sample is adjudged to contain few, if any, ultrafine particles and fewer numbers of fine particles compared with the other samples. This suggests that quartz, which also has the lowest surface area of the samples, exhibits high bioreactivity by virtue of its crystalline silicon dioxide surface. Other workers (Wolff *et al.*, 1988) have emphasized the aggressive bioreactivity of crystalline gallium oxide at a lung burden of 0.8 mg in the rat.

The second silicon dioxide, amorphous Cabosil, does have a large reactive surface area together with a large number of ultrafine and fine particles (*Figure 7.5*). Cabosil produces increases in lung permeability,

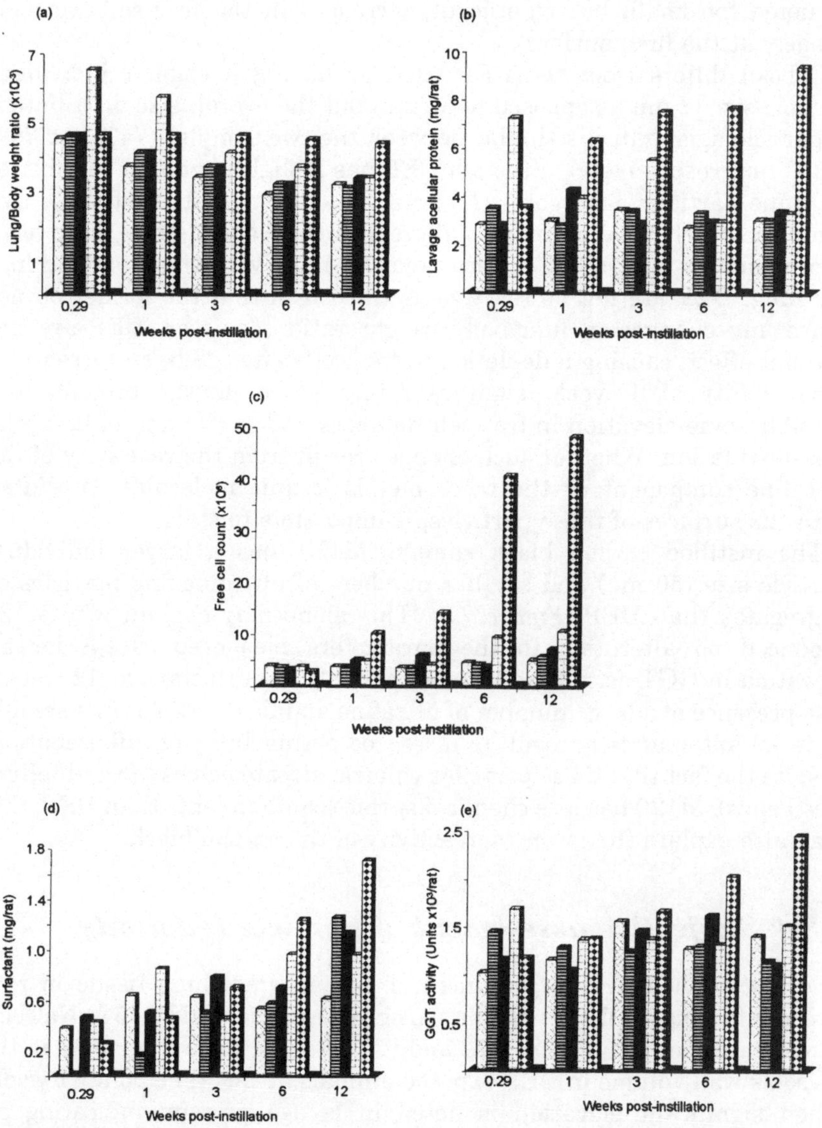

Figure 7.6. Measurement of (a) lung/body weight ratio; (b) lavage protein (1000 ***g*** supernatant); (c) free cell number; (d) lavage surfactant; (e) lavage gamma glutaryl transpeptidase (GGT) activity (300 ***g*** supernatant) in rats instilled with M120 carbon black ▤, diesel exhaust particulates ■, Cabosil ⊞, DQ12 quartz ⁙ or carrier only ▨ at different time points post-instillation.

neutrophils in lavage (not shown) and GGT in lavage indicating epithelial damage at 48 h post-instillation. The effects however, do not persist and lung/body weight, protein lavage levels and GGT activity are gradually restored to that observed in untreated rats. The residual effect of Cabosil at 6–12 weeks post-instillation is a mild (non-neutrophil) inflammation, as noted by small, but significant, increases in the free cells (macrophages) at the lung surface.

Cabosil differs from the DEP used by having a smaller individual particle size (7 nm as opposed to 30 nm) but the overall size distribution of particle aggregates is similar between the two samples (74% and 79% <300 nm, respectively), although DEP has a slightly greater number of ultrafine particles/aggregates (*Figure 7.5*). This suggests that these two samples, which have distinctly different surface chemistries, may have approximately equivalent surface areas for reactivity when instilled into the lung. DEP are not as reactive as Cabosil in that the former do not cause any changes in lung/body weight ratio. However, DEP are not without effect, causing a depletion in free cells after 48 h, an increase in permeability at 1 week (significant increase in lavage protein) and possibly some elevation in free cell numbers and surfactant at 12 weeks post-instillation. Whether such changes result from the reactivity of the ultrafine component or the trace metals/organic molecules associated with the surfaces of these particles, is impossible to state.

The instilled carbon black sample, M120, has a larger individual particle size (50 nm) and smaller numbers of ultrafine/fine particles or aggregates than DEP (*Figure 7.5*). This alone may explain why M120 produced no alteration in the parameters measured except for an elevation in GGT activity (48 h) and possibly in surfactant at 12 weeks. The presence of a large number of ultrafine and fine particles in a sample does not automatically result in increased permeability or inflammation, despite the fact that these particles will translocate across the epithelium (see below). M120 has less chemical surface contaminents than DEP, this may also explain the lower bioreactivity of the carbon black.

7.3.4 Subjective assessment of particle reactivity

The histopathological appearances of parenchymal lung tissue in rats exposed to 1 mg of DEP by instillation or to the carrier fluid 0.15 M NaCl for 1 week are shown in *Figure 7.7a* and *b*. There is a general increase in the alveolar wall volume in DEP-exposed animals at this time point (1 week) when significant elevation is noted in lavage protein, indicating an increase in lung permeability (*Figure 7.6b*). The changes noted from histopathology are transient and substantially reduced at 3 weeks post-instillation. At 6–12 weeks post-instillation lung tissue from control and DEP-treated rats have identical architecture with the exception of occasional thickened areas of alveolar septa (*Figure 7.7c*). Comparative histopathological studies with the other particles (carbon black, Cabosil and quartz) were not carried out in this study but DEP, administered at this small dose, clearly do not produce severe inflammation or extensive thickening of the epithelium as described previously for quartz (*Figure 7.4b* and *c*).

A second subjective guide to bioreactivity of the particles is to examine the lymph nodes associated with the lung, as these increase in size

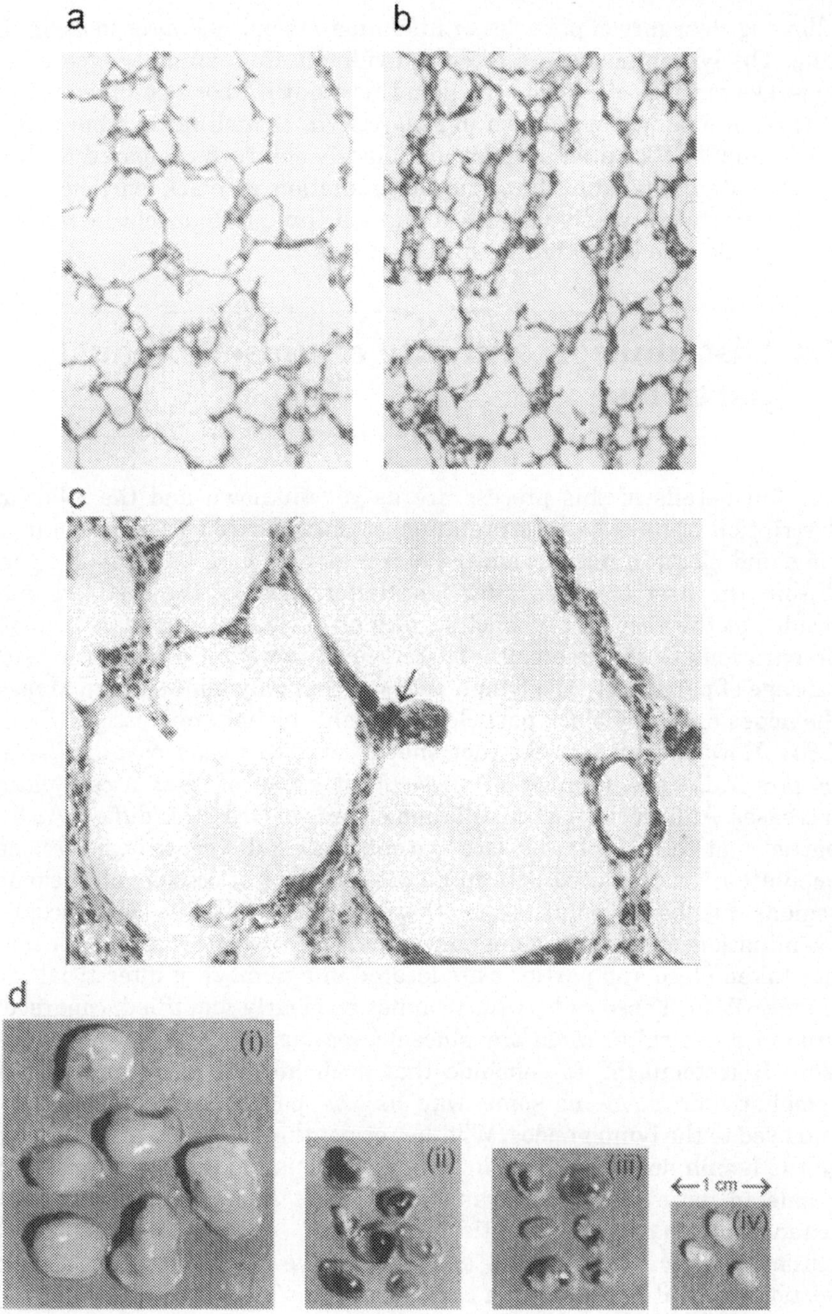

Figure 7.7. Light micrographs of rat parenchyma and lymph nodes. (a) One week after instillation of carrier fluid or (b) diesel exhaust particulates (DEP) (1 mg). Note increase in the volume of the alveolar walls in (b). (c) Occasional alveolar septa (arrow) become thickened at 6 weeks post-instillation of 1 mg of DEP. (d) Lymph nodes from (i) quartz, (ii) DEP, (iii) M120 carbon black and (iv) carrier only-treated rats at 30 weeks post-instillation of 1 mg of dusts. Note dark collections of particles in focal areas for DEP and M120. Quartz-treated animals show extensive increases in lymph node size compared to carrier only-treated animals.

following clearance of particles or inflammation/epithelial response in the lung. The lymph nodes, dissected individually from animals exposed for 30 weeks to three of the four dusts in this study (Cabosil not investigated at this time point), are shown in *Figure 7.7d.* Lymph nodes from carbon black- and DEP-treated animals are slightly enlarged compared to those of untreated rats and show the accumulation of black particles. The progressively bioreactive quartz induces a three to four-fold increase in lymph node size compared to untreated rats.

7.4 Distribution and fate of diesel exhaust particles

The full details of this process are as yet unknown and the following description applies to interpretation of data derived from preliminary light and electron microscopical examination of lung and lymph nodes. Within the first 3 weeks after instillation most of the instilled mass resides at the respiratory tract surface or is clearly seen in centrifuged preparations of the free cells. There seems therefore to be a very slow passage of particles to the lymph nodes in that only minute, focal areas of the nodes have any black particles detectable by light microscopy (*Figure 7.8a*). However, at 6 weeks (not shown) and 12 weeks post-instillation (*Figure 7.8b*) the number of these focal areas of dust accumulation increases. At 6 weeks post-instillation of 1 mg of DEP, it is difficult to find particles at the alveolar surface (except in the total centrifuged free cell population) or deposited within the tissue except at occasional thickened regions of the alveolar septa (*Figure 7.7c*). Electron microscopical examination of these regions show that translocation of the particles has taken place and particles are located in a number of interstitial cells (*Figure 7.8c*). These cells, which cannot be clearly identified, congregate around a central electron translucent area, again of unknown composition. It is tempting to conclude that such an area is connected with lymphatic drainage in some way as the particles are subsequently removed to the lymph nodes. Whether or not the particles are transported to the lymph nodes by these interstitial cells is unknown. The cells that reside in, or overlay, the translucent areas appear to show signs of extensive damage being particularly devoid of organelles and nuclear chromatin. From preliminary examination we have found that a single section of an alveolar septum contains some 4000–5000 individual DEP at 6 weeks post-instillation, thus, indicating that the particles are aggregating in very focal areas of the lung tissue. This aggregation, but on a much larger scale, continues in the lymph nodes. At 24 weeks post-instillation, some nodes have massive accumulations of DEP located in the hilar regions (*Figure 7.8d*) and these are the areas that can be seen by naked eye as described previously (*Figure 7.7d*).

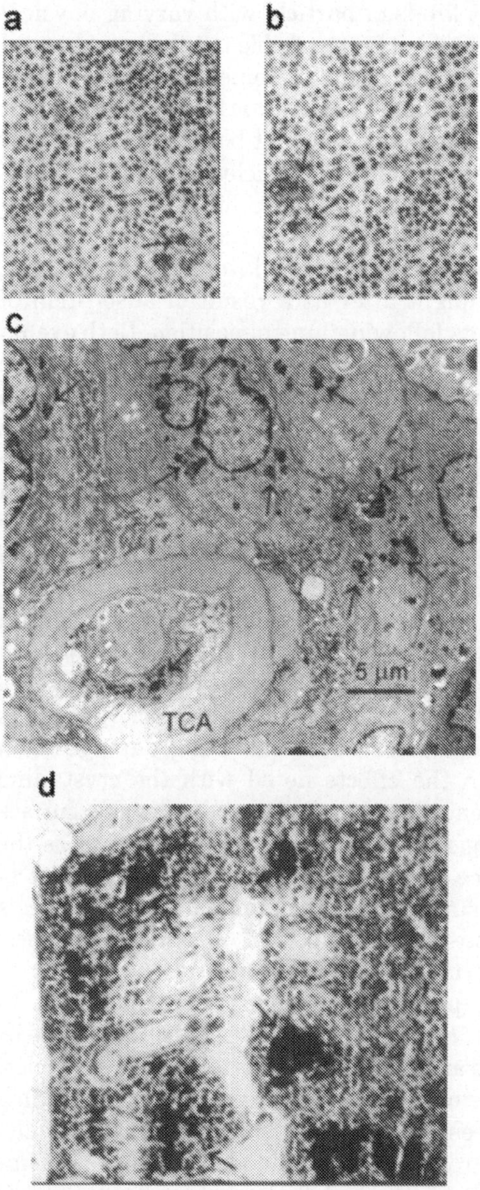

Figure 7.8. Localization of instilled diesel exhaust particulates (DEP) (1 mg) at different time periods after instillation (arrows). (a) Small, focal areas of particles at 1 week post-instillation. (b) Focal areas of DEP more readily detected in lymph nodes after 12 weeks. (c) Electron micrograph of part of an alveolar septum 6 weeks post-instillation. A number of interstitial cells containing aggregated DEP surround an electron translucent central area (TCA) overlaid by a dead or dying cell which also contains particles. (d) Large aggregates of DEP in hilar regions of the lymph nodes at 24 weeks post-instillation.

7.5 Summary and comment

The sequential cellular alterations produced in the lung following toxicant challenge as they relate to primary damage, permeability changes, inflammation and epithelial repair have been described. Quantitative assessment methods were used to compare and contrast the bioreactivity of four different kinds of particle with varying physical and chemical properties. Unfortunately, it is impossible to instill an identical size distribution of particles of differing chemical composition to elucidate whether individual particle size or surface chemistry is the most important feature to explain bioreactivity. Logic dictates that both are important.

An ultrafine sample will contain billions of particles with a huge total surface area for reaction with the lining fluid and cells of the respiratory tract, and thus, the smaller the particle the greater the potential reactive surface area. However, ultrafine particle/particle interactions may take place or the surface properties might alter as a result of adsorption of epithelium lining fluid components following lung deposition. Both events would increase the likelihood of particle aggregation. This helps reduce the available reactive surface area of a sample of particles, to localize the insult and any resulting lesion; an effect noted with the weakly bioreactive DEP used in this study. By contrast, quartz at the same mass dose as DEP, without the presence of any ultrafine particles, a much reduced surface area for interaction and little tendancy for aggregation, has the capacity to produce progressive lung damage which is also reflected in lymph node enlargement. Thus, particle size seems less important with quartz, but rather the chemical reactivity of silicon dioxide and the crystalline nature of this mineral (as also noted with crystalline gallium oxide) dictates its activity. Cabosil, which has a large ultrafine component, a potentially large surface area for reactivity and is pure amorphous silicon dioxide induces immediate lung damage upon instillation but does not sustain the effects noted with the crystalline quartz. This again provides evidence that small size and extensive surface area, especially if the particles aggregate, may not be as important as the chemical reactivity of the surface of the particle in the development of progressive lung damage. This is supported by the minimal biological effects of the carbon black, M120, that has large numbers of ultrafine/fine particle aggregates but a relatively unreactive oxidized carbon coat.

The studies with M120, in particular, suggest that increases in permeability, inflammation or epithelial demise/repair may not be necessary for the epithelial translocation of ultrafine/fine particles. Translocation appears to be a slow process, but is substantial, leading primarily to interstitialization especially in alveolar septa and finally large accumulations in the hilar regions of lymph nodes. The mechanisms by which small masses of particles with weak biological reactivity trigger this process are still unknown.

Acknowledgements

We would like to thank the Medical Research Council, UK and the Department of Environment, Transport and the Regions for financial support for some of this work.

References

BéruBé, K.A., Jones, T.P., Williamson, B.J., Winters, C., Morgan, A.J. and Richards, R.J. (1999a) Physicochemical characterization of diesel exhaust particles: factors for assessing biological activity. *Atmos. Environ.* **33** (10): 1599–1614.

BéruBé, K.A., Jones, T.P., Williamson, B.J. and Richards, R.J. (1999b) The physicochemical characterization of urban airborne particulate matter. In: *Particulate Matter* (eds R.L. Maynard and C.V. Howard). BIOS Scientific Publishers, Oxford, pp. 39–62.

Crouch, E.C., Moxley, M.A. and Longmore, L. (1987) Synthesis of collagenous proteins by pulmonary type II epithelial cells. *Am. Rev. Respir. Dis.* **135**: 1118–1123.

Devereux, T.R., Serabjit-Singh, C.J., Slaughter, S.R., Wolf, C.R., Philpot, R.M. and

Dinsdale, D. (1998) Biotransformation in the pulmonary epithelium: the role of different cell types. *Resp. Med.* **92** (11): A10.

Dinsdale, D., Green, J.A., Manson, M.M. and Lee, M.J. (1992) The ultrastructural immunolocalization of gamma-glutamyl transpeptidase in rat lung: correlation with the histochemical demonstration of enzyme activity. *Histochem. J.* **24**: 144–152.

Fouts, J.R. (1981) Identification of cytochrome P-450 isozymes in non-ciliated bronchiolar epithelial (Clara) and alveolar type II cells isolated from rabbit lung. *Exp. Lung Res.*, **2**: 221–230.

Green, G.M. and Watson, A.Y. (1995) Relation between exposure to diesel emissions and dose to the lung. In: *Diesel Exhaust: A Critical Analysis of Emissions, Exposure and Health Effects.* Health Effects Institute, Cambridge, MA, pp. 167–184.

Hooftman, R.N., Kuper, C.F. and Appleman, L.M. (1988) Comparative sensitivity of histopathology and specific lung parameters in the detection of lung injury. *J. Appl. Toxicol.* **8**: 59–65.

Horton, J.K., Meredith, M.J. and Bend, J.R. (1987) Glutathione biosynthesis from sulphur-containing amino acids in enriched populations of Clara and type II cells and macrophages freshly isolated from rabbit lung. *J. Pharmacol. Exp. Ther.* **240** (2): 376–380.

Masek, L. (1990) Correlation of Biochemical and Histopathological Assessment of Lung Damage. MPhil. Thesis, University of Wales.

McElroy, M.C., Pittet, J-F., Allen, L., Wiener-Kronish, J. P. and Dobbs, L.G. (1997) Biochemical detection of type I cell damage after nitrogen dioxide-induced lung injury in rats. *Am. J. Physiol.* **273**: L1228–L1234.

Miller, B.E., Dethloff L.A. and Hook, G.A. (1986) Silica-induced hypertrophy of type II cells in the lungs of rats. *Lab. Invest.* **55** (2): 153–163.

Miller, Y.E., Walker, S.R., Spencer, J.S., Kubo, R.T. and Mason, R.J. (1989) Monoclonal antibodies specific for antigens expressed by type II alveolar epithelial and non-ciliated bronchiolar cells. *Exp. Lung Res.* **15**: 635–649.

Murphy, S.A., BéruBé, K.A., Pooley, F.D. and Richards, R.J. (1998) The response of lung epithelium to well characterised fine particles. *Life Sci.* **62** (19): 1789–1799.

Oberdörster, G., Ferin, J., Gelein, R., Soderholm, S.C. and Finkelstein, J. (1992) Role of the alveolar macrophage in lung injury: studies with ultrafine particles. *Environ. Health Perspect.* **97**: 193–199.

Richards, R.J. and Curtis, C.G. (1984) Biochemical and cellular mechanisms of dust-induced lung fibrosis. *Environ. Health Perspect.*, **55**: 393–416.

Richards, R.J., Masek, L.C. and Brown, R.F.R. (1991) Biochemical and cellular mechanisms of pulmonary fibrosis. *Toxicol. Path.* **19** (4): 526–539.

Sage, H., Farin, F.M., Striker, G.E. and Fisher, A.B. (1983) Granular pneumocytes in primary culture secrete several major components of the extracellular matrix. *Biochemistry* **22**(9): 2148–2155.

Serabjit-Singh, C.J., Nishio, S.J., Philpot, R.M. and Plopper, C.G. (1987) The distribution of cytochrome P-450 monooxygenase in cells of the rabbit lung: an ultrastructural immunocytochemical characterization. *Mol. Pharmacol.* **33**: 279–289.

Wolff, R.K., Henderson, R.F., Eidson, A.F., Pickrell, J.A., Rothenberg, S.J. and Hahn, F.F. (1988) Toxicity of gallium oxide particles following a 4-week inhalation exposure. *J. Appl. Toxicol.* **8** (3): 191–199.

8 The toxicology of ultrafine particles

Ken Donaldson, Vicki Stone and William MacNee

8.1 The lung and particle defence

Respirable particles have always been present in the environmental air to some extent, from wind-blown dust, fire *et cetera* and the mammalian lung has evolved to deal with them. Environmental sources such as wind-blown dust and fires tend to produce particles that range down to about 0.5 μm in diameter. Particles in this size range that deposit in the distal airways and centriacinar region are often described as 'fine' in contrast to 'ultrafine' particles which are particles less than 0.1 μm (100 nm) (Dy and Chen, 1997) and which are described later. The airways are well protected against deposited fine particles by the muco-ciliary escalator which is effective in trapping deposited particles and transporting them up and out of the lungs. Moving peripherally however, towards the terminal airways and the alveoli, the muco-ciliary escalator becomes less effective and clearance via that route is subsequently slower (Schlesinger *et al.*, 1998). In the terminal airways and alveolar region the macrophages take over as the dominant clearance mechanism. Macrophages phagocytose fine particles and can then take one of several routes that depend on the characteristics of the particle and the exposure. The different toxicological outcomes and particle clearance pathways for different particle types have been investigated in the rat and are shown in *Table 8.1*. It is important to stress that the relative importance of the different routes of transport are best understood (but still not completely understood) in the rat. However, particle clearance and retention in the distal lung is not the same in the

Particulate Matter, edited by R.L. Maynard and C.V. Howard

Table 8.1. Toxicological particle exposures and particle movement in the rat

Particle exposure	Macrophage response in rats	Route of particle movement
Non-toxic particles: low exposure (e.g. carbon)	Macrophages phagocytose particles and migrate to MCE leading to effective clearance.	Minimal interstitial and lymph node transfer; most particles retained in the alveolar macrophages or cleared in macrophages via MCE.
Non-toxic particles: high exposure	Overload conditions eventually occur with slowed clearance, inflammation leading to fibrosis and tumours.	Increased interstitial and lymph node transfer and retention in macrophages in the lumen; impaired clearance.
Cytotoxic particles (e.g. quartz)	Macrophages are killed or activated, leading to inflammation, fibrosis and cancer.	Increased interstitial and lymph node transfer; impaired clearance via the macrophages.

MCE, muco-ciliary escalator.

rat as in other species of toxicological importance such as the mouse, hamster, primates and of course man (Nikula *et al.*, 1997).

8.2 Ultrafine particles

Ultrafine particles are particles less than 100 nm in diameter, such as ultrafine carbon black, ultrafine TO_2 and ultrafine metals that are produced industrially (see Dy and Chen, 1997). Some occupational exposure to these substances is to be anticipated and so a knowledge of their toxicology is needed.

8.2.1 Ultrafines and PM_{10}

PM_{10} is the common global sampling convention that collects and assesses the mass of particulate air pollution in the environment. The PM_{10} convention collects particles that are centred around 10 μm in aerodynamic diameter (D_{ae}) so PM_{10} may contain a few particles of 15 μm, has 50% efficiency for particles of 10 μm D_{ae} and collects down into the ultrafine range. PM_{10} is derived from multiple sources but traffic-derived pollution is a major source of urban PM_{10} (Maynard and Waller, 1996). Ultrafine particles are generated in vehicle exhaust such as diesel engine exhaust (Kittelson, 1998) and these are an important contributor to the environmental particulate air pollution (see below). There is ample evidence that PM_{10} contains an ultrafine component that is derived from combustion sources such as diesel (Oberdörster *et al.*, 1995; Peters *et al.*, 1997). Diesel exhaust is comprised of singlet particles equal to or less than 50 nm diameter (Maynard and Waller, 1996) which tend to form aggregates soon after generation if they are at sufficient concentration.

Several studies on PM_{10} have described a substantial component in number terms, of particles in the ultrafine range (e.g. Maynard and Waller, 1996; Peters *et al.*, 1997) although this represents a relatively small fraction of the total mass (Peters *et al.*, 1997).

Compared to the gaseous components, such as ozone and NO_x, PM_{10} has been found to be the most harmful of the common air pollutants causing a substantial number of asthma and COPD attacks and hospitalizations as well as death from respiratory and cardiovascular cause (e.g. Pope *et al.*, 1995). It is to be noted that these adverse health effects occur in a small sub-section of the population that are susceptible. PM_{10} is present at a very low mass concentration in urban air, commonly 10–50 μg m^{-3}, and this means that some component of PM_{10} is likely to be highly toxic in the susceptible populations. Acute effects of PM_{10} have been demonstrated in the hours and days following air pollution episodes and chronic effects also occur (Dockery *et al.*, 1993).

8.2.2 Adverse health effects of particles

The best studied particles in terms of their adverse health effects are those that are encountered in the occupational setting. Examples of the best studied particles and their adverse health effects are shown in *Table 8.2*.

As can be seen from *Table 8.2* only PM_{10} and welding fumes appear to have acute effects whilst the majority of particles have chronic effects. Welding fume particles, like PM_{10}, are also very small (Fine *et al.*, 1997) and this may help to explain the unusual acute nature of the effects seen with welding fumes and PM_{10}. Thus PM_{10} appears to act differently to the common pathogenic particles that are encountered in occupational settings and this may be a consequence of the ultrafine component. PM_{10} also has chronic effects (Dockery *et al.*, 1993) but these are more difficult to study toxicologically although of possibly great importance.

We have advanced a hypothesis to explain the adverse health effects of PM_{10} based on the toxicity of the ultrafine particle component (Seaton *et al.*, 1995). We have further refined this hypothsis (Donaldson and MacNee, 1998) to suggest that generation of free radicals at the particle surface may lead to inflammation in the lung; this local inflammation

Table 8.2. The different effects of the best known pathogenic particles in relation to PM_{10}

Particle type	Concentration at which adverse effects are seen	Time-scale of the effect	Effect
PM_{10}	10–50 μg m^{-3}	Hours, years	Asthma and COPD exacerbations; death from respiratory causes, heart attack, stroke
Asbestos	<100 fibres per ml	Years	Asbestosis, lung cancer, mesothelioma
Quartz	<1 mg m^{-3}	Years	Silicosis, cancer
Coalmine dust	<10 mg m^{-3}	Years	Pneumoconiosis
Welding fumes	<1 mg m^{-3}	Hours	Fever etc.

may lead to enhanced oxidative stress and culminate in a pro-coagulant state that favours heart attacks and strokes in susceptible populations.

The types of sample that are available make it difficult to define the role, if any, of ultrafines in the health effects of PM_{10}. As shown in *Table 8.3* the samples that are available are from three sources and none of them match with airborne PM_{10}. Most work on the mechanism of the effects of PM_{10} has been on filter-derived samples and these are likely to be quite variable with regard to the yield of aggregates of ultrafines. At the other extreme are 'pure' ultrafine preparations which comprise particles in a very tight size distribution of ultrafine range, such as ultrafine carbon black (see later). These are aggregates of variable size comprising primary ultrafine particles; these can have very low metal contents. Therefore these are not directly comparable to PM_{10}. In the USA, particle concentrators have been used to increase the mass concentration in the air that is piped directly to laboratory animals whose responses are measured (e.g. Sioutas *et al.*, 1997). Such devices do not raise the concentration of the ultrafine fraction and so do not perfectly represent a concentrated ambient aerosol.

8.3 Toxicology of ultrafine particles

8.3.1 Deposition

Ultrafine particles deposit with high efficiency in the respiratory tract (Cheng and Swift, 1995; Yeh *et al.*, 1997). In COPD the deposition of ultrafines has been reported to be more efficient due probably to changes in the characteristics of air flow within the lungs (Anderson *et al.*, 1990).

8.3.2 Human studies

Peters *et al.* (1997) reported that decrements in evening peak flow in a group of asthmatics was associated with the ultrafine component of the airborne particles during pollution episodes. The study demonstrated the best association with the ultrafine fraction, although there were associations with other size fractions. These studies of ambient environ-

Table 8.3. Comparison of the components of airborne PM_{10} versus the particle samples available for toxicological study

Particulate	Singlet ultrafines	Aggregates of ultrafines	Transition metals
Airborne PM_{10}	✓	✓	✓
Filter samples	No	✓	✓
Particle concentrators	✓ but not concentrated	✓	✓
Ultrafine carbon black	No	✓	No

mental exposures attempt to demonstrate an effect of ultrafine particles within a mixed exposure that includes other pollutants. Chamber studies allow a single, controlled exposure and Kuschner *et al.* (1997b) exposed human volunteers for up to 30 min to 33 mg m^{-3} ultrafine zinc oxide particles, average primary particle size 8–40 nm. Large increases in the bronchoalveolar lavage (BAL) concentrations of the pro-inflammatory cytokines tumour necrosis factor α (TNFα) and interleukin (IL)-8 were found 3 h after cessation of exposure. However, in a similar study exposing subjects to magnesium oxide particles where 28% of particles were < 100 nm in diameter, there was no effect on the cytokine levels in the lavage (Kuschner *et al.*, 1997a). The dosimetry of these two studies was not identical and so it is difficult to compare them but toxicological studies on the two particle types might be able to determine why one ultrafine type was inflammogenic and one was not.

8.3.3 Animal studies

Overload and the MTD. Because of the increasing elucidation of the phenomenon of overload in laboratory rats exposed to high levels of low-toxicity dusts, there needs to be careful interpretation of particle toxicology studies. The maximum tolerated dose (MTD) is an important concept in this regard. 'Classical overload' occurs in rats exposed to high levels of airborne particles which accumulate in the lungs to a point where there is failure of clearance, increased build-up of dose, inflammation, proliferation, fibrosis and tumour production (e.g. see papers in Mauderly and McCunney, 1996). Therefore consideration needs to be given to the possibility that a sufficiently high exposure to any airborne particle, even a 'low-toxicity particle' that is non-pathogenic at plausible levels of human exposure, will result in a false-positive pathological outcome. The MTD, as it applies to particle toxicology, is a concept that takes the possibility of overload-related pathology into account and seeks to carry out the exposure at a level that does not cause overload. The MTD of a 'low-toxicity' particle is the maximum dose that dose not cause a slowing down of clearance.

This is particularly important in the case of ultrafine particles since the dose metric that seems to predict overload is not mass but surface area of particles in the lungs (Tran *et al.*, 1998). A given mass of ultrafine particles therefore has a much greater surface area than the same mass of fine, yet respirable, particles. This is reflected in data showing that for a range of non-toxic particles, lung burden expressed as surface area predicts inflammogenicity (Oberdörster, 1996) and tumorigenicity (Driscoll, 1996). Thus, from a toxicological viewpoint, it is important to separate the effects of ultrafine particles at overload from effects at non-overload, if they occur. Further complications arise when it is appreciated that overload may be a phenomenon confined to rats (e.g. Donaldson, in press).

The proportion of PM_{10} that is ultrafine is very small in mass terms, unlikely to be more than a few μg per m^3 (Peters *et al.*, 1997). If the ultrafine component is important for the adverse health effects, it does not cause these effects in humans by an overload mechanism because this is a phenomenon that only occurs at high exposure concentration and, even then, probably only in rats.

8.3.4 Inhalation studies with ultrafine particles at overload

Ferin *et al.* (1992) demonstrated that when rats were exposed to an equal airborne mass concentration of fine and ultrafine TiO_2 particles, there was far more bronchoalveolar inflammation in the case of the ultrafine. Agglomeration meant that the aerodynamic diameter of the ultrafine particle cloud was close to that of the fine cloud, ensuring similar deposition in the two experimental groups. In tandem with the increased inflammation, these rats also showed a notable increase in interstitial and lymph node transfer of the ultrafines compared to the fines. The airborne mass concentration of 23 mg m^{-3} was clearly an overload exposure. Driscoll *et al.* (1996) demonstrated inflammation and mutation in the lungs of rats exposed to ultrafine carbon black at overload exposure levels. A number of studies at overload exposure levels have demonstrated overload pathology in rats with the ultrafine forms of carbon black and TiO_2 (Heinrich *et al.*, 1995).

8.3.5 Non-overload studies

Johnson *et al.* (1996) reported that very low airborne mass concentrations of ultrafine particles of polytetrafluoroethylene (PTFE) caused a rapid and severe neutrophil influx that is associated with the increased transcription of anti-oxidant genes such as metallothionein and Mn-SOD (Mn-superoxide dismutase), and also cytokines. Oxidative stress, leading to the transcription of oxidative stress-responsive genes such as Mn-SOD could be a result of activation of inflammatory phagocytes but they could also result from free oxygen and fluoride radicals on the surface of the PTFE fume (Johnston *et al.*, 1996). The latter possibility means that PTFE fume may not be considered a low-toxicity particle and is in the category of toxic dusts, where an explanation of toxicity in terms of size becomes clouded by the issue of direct toxic effects of the surface. The concept of overload is confined to non-toxic dusts.

There are few studies on the effects of standard ultrafine particles of low toxicity in the non-overload situation. In order to try and better understand the effects of ultrafine particles at low airborne mass concentration we have examined the effects of ultrafine carbon black (ufCB 14 nm primary particle size) and non-ultrafine carbon black (CB 260 nm average primary particle size) at 1 mg m^{-3} for 7 h in rats. This exposure had no effect in the rats exposed to CB but the ufCB caused several pro-inflammatory effects (*Table 8.4*).

Table 8.4. Summary of the changes in bronchoalveolar lavage (BAL) lung tissue and plasma indices following 7 h inhalation exposure to 1 mg m^{-3} fine carbon black (CB) or ultrafine carbon black (ufCB)

BAL parameter	CB	ufCB
BAL PMN	No effect	Increase
BAL protein	No effect	Increase
Lung glutathione	No effect	Decrease
Plasma antioxidants	No effect	Decrease
Plasma Factor VII	No effect	Increase

Data summarized from MacNee *et al.* (submitted).

In addition to local effects in the lung, systemic changes were found with ufCB only. There was evidence of oxidative stress as measured by the TEAC assay (Rahman *et al.*, 1996) and increased plasma Factor VII. The latter result is particularly interesting since increased Factor VII is an independent risk factor for cardiovascular disease (Catto and Grant, 1995).

These data demonstrate both a local inflammatory effect and a systemic effect of ultrafine particles at low dose. These data need to be interpreted back to the human exposure scenario with caution. However, it does mark a clear link between low exposure to ultrafine particles, known to be present in PM_{10} pollution, and physiological changes relevant to pulmonary inflammation, heart attacks and strokes as advanced in our original hypotheses (Seaton *et al.*, 1996).

8.3.6 Instillation studies

The technique of instillation is used as an alternative to inhalation as a mode of exposure. This technique has to be interpreted with caution since it is characterized by non-physiological delivery, high dose and dose-rate. Li *et al.* (1996) utilized the ultrafine and fine carbon blacks described above and compared the inflammation produced by instillation of an equal mass of 125 μg. There was more inflammation and oxidative stress caused by the ufCB. However, in follow-up studies, where the dose was increased, more inflammation was produced by the CB than ufCB at the higher dose (Li *et al.*, in press). This can be interpreted as a weakness of the instillation model where there is marked interstitial transfer of the ultrafines. This may cause a change in the chemotactic gradient towards the interstitium, resulting in a misleading decrease of leukocytes available for BAL inflammatory indices at higher doses (Oberdörster *et al.*, 1994).

The extent of inflammation following instillation discriminated well between ultrafine and non-ultrafine Ni (Zhang *et al.*, 1998b) with much more inflammation caused by the ufNi. The ufNi particles generated large quantities of hydroxyl radicals quantified by the ability to cause scission of plasmid DNA (Zhang *et al.*, 1998a), a sensitive measure of hydroxyl radical activity. It is therefore informative to determine whether increased surface reactivity could be an alternative or additional factor in the greater inflammogenicity of any ultrafine sample, compared to its non-ultrafine alternative (*Figure 8.1*). Metals are a special case of toxic particles and metals like Ni have well-documented toxicity and so it is to be anticipated that a greater surface area of metal might mediate greater toxicity to the lungs. Metals are a component of PM_{10} and welding fumes and so they may provide a mechanism for lung inflammation. If the metals were in the ultrafine form they could be potentially highly inflammogenic.

8.3.7 In vitro studies on ultrafine particles

There have been some studies on the *in vitro* pro-inflammatory effects of ambient particles and in general these have shown the effects to be mediated by the transition metal content. However, because the extent of retrieval of ultrafine particles and aggregates is unknown, the role of ultrafines in these filter-derived samples must be questioned (Carter *et al.*, 1997; Samet *et al.*, 1996).

There have been few studies on the *in vitro* effects of ultrafine particles without transition metal contamination. In the case of ufCB there is no evidence that there is an important contribution from transition metals and indeed there is no difference in the inflammation caused by instilled ufCB compared to ufCB pre-treated with iron chelator desferrioxamine nor any biological activity in leachates from ufCB (Brown *et al.*, in preparation). We have demonstrated that ufCB is capable of causing oxidative stress to epithelial cells in culture, as shown by glutathione depletion and that this is related to greater free radical activity associated with the surface of the ufCB (Stone *et al.*, 1998); CB was much less active in these respects. This does not appear to be related to transition metal content since leachates collected from ufCB that would contain the soluble transition metals, do not cause lung inflammation. PM_{10} collected in Edinburgh has free radical activity that is generated by Fenton chemistry from Fe^{3+} (Gilmour *et al.*, 1996) but we have no evidence that this Fe^{3+} is derived from the ultrafine component, although all the free radical activity could be recovered in a supernatant when the particles were pelleted by centrifugation (Gilmour *et al.*, 1996). However, there is greater free radical activity at the surface of ufCB than CB when they are compared on an equal mass basis (Stone *et al.*, 1998).

A mechanism for the increased inflammation caused by ufCB *in vivo* was suggested by studies on the Ca^{2+} levels in cells exposed to ufCB and CB (Stone *et al.*, submitted). Using the fluorescent dye Fura 2, which detects intracellular Ca^{2+}, we compared the effect of ufCB and CB on the cytoplasmic Ca^{2+} levels of a human macrophage cell line, Monomac 6 (MM6). This showed there to be a rapid (within minutes) small increase in the resting Ca^{2+} current and a very large increase in the entry of Ca^{2+} when the cells were stimulated with thapsigargin, an agent known to cause release of Ca^{2+} stores from the endoplasmic reticulum (Stone *et al.*, submitted) leading to activation of the calcium release-activated calcium current and hence entry of extracellular calcium. This is interpreted as a rapid direct effect of the ultrafine particles on the membrane Ca^{2+}

channels allowing greater entry of extracellular Ca^{2+} in response to stimulation. This was supported by the fact that the effect of the ufCB on the cytoplasmic Ca^{2+} concentration could be blocked with verapamil, the Ca^{2+} channel blocker. Further evidence was produced by repeating the experiments in Ca^{2+}-free medium and in the presence of Mn^{2+} which can only enter the cells via the Ca^{2+} channels. Treatment of MM6 with ufCB in the presence of Mn^{2+} led to a decrease of Fura2 fluorescence indicative of Mn^{2+} entry via the opened Ca^{2+} channels. Effects of residual oil fly ash (ROFA) particles on intracellular calcium concentrations in human bronchial epithelial cell lines have also been observed (Veronesi *et al.*, 1999). In this cell line increased intracellular Ca^{2+} levels stimulated with ROFA have been associated with increase pro-inflammatory cytokine gene expression, but the role of ultrafines is unknown.

8.3.8 The cellular/molecular mechanism of ultrafine particle toxicity

Ultrafine particles may initiate inflammation because:

- Their small physical size means that there are extremely large numbers of particles per unit mass; the sheer numbers of particles depositing may forestall their efficient phagocytosis and clearance via the macrophage system, allowing the particles to interact with the epithelial cell surface causing Ca^{2+} effects via direct interaction with the plasma membrane Ca^{2+} channels. This could lead to the production of pro-inflammatory cytokines being transcribed in response to the activation of transcription factors such as NF-κB.
- Their small size may present an increased surface area for generation of free radicals as a result of redox activity at the particle surface that is transferred to the cell interior resulting in oxidative stress and activation of transcription factors for pro-inflammatory proteins.
- Their small size may present an increased surface area for release of transition metals to generate oxidative stress and cause activation of pro-inflammatory transcription factors.

There is no reason why these should be mutually exclusive and there could be a different contribution of the two pathways in the pathogenic mechanism of any ultrafine particle sample. For instance, there is good evidence of a role for metals in the toxicity of ambient particles of ROFA (Dreher *et al.*, 1995), fly ashes (Pritchard *et al.*, 1996) and PM_{10} (Gilmour *et al.*, 1996). All of these samples were, however, retrieved from filters and it is difficult to determine the efficiency of recovery of ultrafine particles which would undoubtedly be present as aggregates since the average particle size in these samples was above 1 μm. It is therefore difficult to know the contribution of the ultrafine components in these samples.

Some light is shed on the role of size by a report that the instillation of 1 mg of 50 nm CB (M120) caused only modest inflammation as measured by neutrophil influx at 2 days, although with no residual inflammation at

1 week or any time thereafter up to 12 weeks (Murphy *et al.*, 1998). The possibility that there is a size cut-off for the biological activity of ultrafines needs to be investigated. In our hands the 50 nm carbon black was substantially less inflammogenic than 14 nm carbon black and more inflammogenic than 260 nm carbon black (*Table 8.5*)

We have also utilized polystyrene particles of various sizes down to 60 nm primary particle size and found no inflammation 24 h after instilling 125 μg (unpublished data), the same mass dose that caused inflammation with ultrafine carbon black. At an instilled mass of 1 mg, however, there was inflammation with the 60 nm polystyrene that was markedly greater in magnitude than was seen with the larger particles. In the light of the data in *Table 8.1*, it would seem that size is important for any specific material but that different materials differ in their relative potential to cause inflammation at any particular particle size.

In keeping with evidence from other pathogenic particles such as quartz (Donaldson and Borm, 1998) and fibres (Brown *et al.*, 1998), the composition of the ultrafine particle, especially its surface characteristics, are likely to be important in mediating pathogenicity.

In a study of three ultrafine metals, Ni, Co and TiO_2, there were differences in the inflammogenicity of the three ultrafines as measured by BAL and this could be related to the ability to cause scission of plasmid DNA *in vitro* (Zhang *et al.*, 1998a). In another study the increased inflammogenicity of ultrafine Ni was found to be related to its ability to cause an increase in the levels of TNFα in the lung-lining fluid (Zhang *et al.*, in preparation). We have studied a range of ultrafine particles and *Table 8.6* shows the relationship between the free radical activity of these particles and their ability to cause inflammation following instillation into the rat lung.

Therefore ultrafine particles that have the ability to generate high levels of free radicals at their surface have inflammogenic activity. It is notable that not all ultrafine particles are equally pathogenic but decreasing the particle size is likely to make any material more inflammogenic (*Figure 8.1*). For example, Murphy *et al.* (1998) used cabosil amorphous silica with a primary particle size of 7 nm and found no inflammation from 2 days to 12 weeks after instillation of 1 mg. The well-documented solubility of amorphous silica which is likely to be enhanced with such a small particle size may be an important factor in preventing toxicity of this material.

Table 8.5. Percentage PMN in lavage 6 h following instillation of 125 mg of carbon black of various primary particle sizes (from Li *et al.*, in press)

Carbon black primary particle size (nm)	Neutrophils in the bronchoalveolar lavage (%) (mean (SEM))
14	39.9 (4.2)
50	13.7 (2.6)
260	4.2 (0.73)

Table 8.6. Relationship between particle characteristics and ability to cause inflammation (after Zhang *et al.*, 1998a)

Ultrafine particle	Inflammogenicity by instillation	Free radical activity	Primary particle size (nm)
TiO_2	+	+	20
Carbon black	+ + +	+ + +	14
Ni	+ + + + +	+ + + + +	20
Co	+ + +	+ +	20

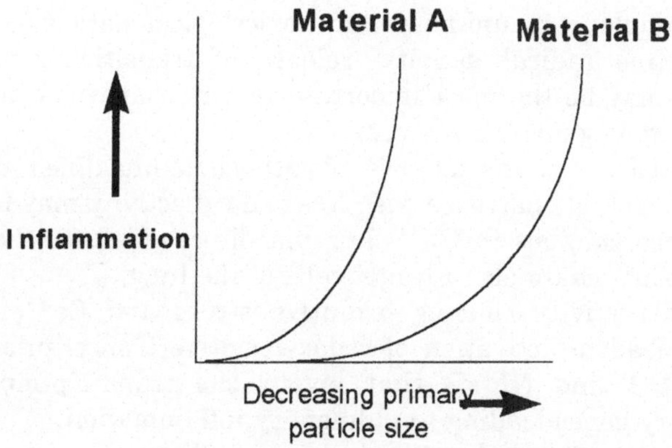

Figure 8.1. Diagrammatic representation of the hypothetical role of size and material in ultrafine particle toxicity.

8.3.9 Hygiene consequences of increased toxicity of ultrafine materials

Currently hygiene standards for dust are based on mass for particulate dusts and fibre number for fibrous dusts. If size is an important factor indicating pathogenicity then expressing the exposure to dusts as *mass* per unit volume of air per unit time may not be a true reflection of the potential of any particle sample to cause lung disease. For some industrial scenarios the measurement of surface area of airborne dust may be an improved index of risk. The provision of a conversion factor for calculation of surface area from mass for a particular exposure or job may suffice.

8.3.10 Conclusions on the toxicology of ultrafine particles

We can conclude the following on the toxicology of ultrafine particles:

- Studies in rats and humans with different types and sizes of ultrafine particles suggest that some ultrafine particles (e.g. CB) have extra toxicity compared to the same material as respirable but non-ultrafine particles when they are present in the lung at the same mass dose.

- This effect may be explained by the greater surface area of the ultrafine material which could deliver oxidative stress because of a greater surface for the release of transition metals or for the generation of free radicals by other means.
- Some ultrafine particles appear to be less or non-pathogenic (e.g. amorphous silica, magnesium oxide); however dosimetric considerations mean that studies with the same particle size in the same species are needed before this can be confirmed.
- The physicochemical factors that dictate the pathogenicity of these ultrafine samples are unknown but toxicological data suggest that smallness, free radical activity, release of transition metals and insolubility may be the most important factors that would make any sample inflammogenic (*Figure 8.2*)
- If the CB model is representative of pathogenic ultrafines in general then small insoluble particles with free radical activity may be able to directly cause changes in Ca^{2+} flux, via direct effect on the plasma membrane Ca^{2+} channels in target cells of the lung.
- Free radical activity causing oxidative stress and Ca^{2+} flux may interact to lead to activation of redox-sensitive transcription factors such as AP-1 and NF-κB that favour the transcription of pro-inflammatory genes leading to pulmonary inflammation.
- Experimental studies with rats suggest that increases in pro-coagulant such as Factor VII, which is an independent risk factor for heart attack and stroke, may occur in individuals exposed to ultrafines at low level for quite short times. Whilst these have to be carefully interpreted in terms of the human response they do show a clear effect of low exposure to ultrafine particles in causing systemic and local pulmonary effects that could impact on heart attacks and strokes.
- Because of the absence of relevant particle samples for studies, the role of ultrafine particles in the adverse health effects of PM_{10} remains unresolved.
- Currently, workplace standards are based on mass measures of the dust per unit volume of air. If surface area is an important metric that drives the inflammatory response of particles then exposure to a given airborne mass concentration of particles in the ultrafine size range carries significantly more risk than the exposure to the same material as fine particles. This has considerable consequences for regulatory bodies.
- More research on the toxicology of ultrafine particles is needed.

Less toxic ⟷	**More toxic**
Increasing size ⟷	Decreasing size
Soluble ⟷	Insoluble
Less free radicals ⟷	More free radicals
Less transition metal ⟷	More transition metals

Figure 8.2. Physicochemical features that may determine the potential toxicity of a sample of ultrafine material.

Acknowledgments

The authors gratefully acknowledge funding from The Medical Research Council, The Colt Foundation and the British Lung Foundation. KD is the Transco British Lung Foundation Fellow in Air Pollution and Respiratory Health. The authors particularly acknowledge the continued and enthusiastic support of Dr Robert Maynard of the UK Department of Health who also read this chapter and made helpful suggestions.

References

Anderson, P.J., Wilson, D.J. and Hirsch, A. (1990) Respiratory tract deposition of ultrafine particles in subjects with obstructive or restrictive lung disease. *Chest* **97**: 1115–1120.

Anderson, D., Blowers, S.D., Marrs, T.C. and Rice, P. (1996) An *in vitro* and an *in vivo* unscheduled DNA synthesis assay with a zinc oxide/hexachloroethane (Zn/HCE) smoke. *Human Exp. Toxicol.* **15**: 38–44.

Becker, S., Soukup, J.M., Gilmour, M.I. and Devlin, R.B. (1996) Stimulation of human and rat alveolar macrophages by urban air particulates: effects on oxidant radical generation and cytokine production. *Toxicol Appl. Pharmacol.* **141**: 637–648.

Brown, D.M., Roberts, N.K. and Donaldson, K. (1998) Effect of coating with lung lining fluid on the ability of fibres to produce a respiratory burst in rat alveolar macrophages. *Toxicol In Vitro* **12**: 15–24.

Carter, J.D., Ghio, A.J., Samet, J.M. and Devlin, R.B. (1997) Cytokine production by human airway epithelial cells after exposure to an air pollution particle is metal-dependent. *Toxicol. Appl. Pharmacol.* **146**: 180–188.

Catto, A.J. and Grant, P.J. (1995) Risk-factors for cerebrovascular-disease and the role of coagulation and fibrinolysis. *Blood Coagulation Fibrinolysis* **6**: 497–510.

Cheng, K.H. and Swift, D.L. (1995) Calculation of total deposition fraction of ultrafine aerosols in human extrathoracic and intrathoracic regions. *Aerosol. Sci. Technol.* **22**: 194–201.

Churg, A., Stevens B. and Wright, J.L. (1998) Comparison of the uptake of fine and ultrafine TiO_2 in a tracheal explant system. *Am. J. Physiol.* **274**: L81–L86.

Dockery, D.W., Pope, C.A., Xu, X.P., Spengler, J.D., Ware, J.H., Fay, M.E., Ferris, B.G. and Speizer, F.E. (1993) An association between air-pollution and mortality in 6 united-states cities. *N. Engl. J. Med.* **329**: 1753–1759.

Donaldson, K. (in press) Critical review of non-neoplastic responses in Proceedings of a Workshop on the relevance of the rat lung response to particle overload for human risk assessment. *Inhalation Toxicol.*

Donaldson, K. and Borm, P. (1998) The quartz hazard: a variable entity. *Ann. Occupat. Hyg.* **42**: 287–294.

Donaldson, K. and MacNee, W. (1998) The mechanism of lung injury caused by PM_{10}. In: *Air Pollution and Health. Issues in Environmental Science and Technology, Number 10* (Eds R.E. Hester and R.M. Harrison), pp. 21–32.

Driscoll, K.E. (1998) Role of inflammation in the development of rat lung tumours in response to chronic particle particle exposure. *Inhalation Toxicol.* **8** (Suppl): 139–153.

Driscoll, K.E., Carter, J.M., Howard, B.W., Hassenbein, D.G., Pepelko, W., Baggs, R.B. and Oberdörster, G. (1996) Pulmonary inflammatory, chemokine, and mutagenic responses in rats after subchronic inhalation of carbon-black. *Toxicol. Appl. Pharmacol.* **136.**

Dy, H.P. and Chen D.R. (1997) Nanometer particles: a new frontier for multidisciplinary research. *J. Aerosol Sci.* **28**: 539–544.

Ferin, J., Oberdörster, G. and Penney, D.P. (1992) Pulmonary retention of ultrafine and fine particles in rats. *Am. J. Respir, Cell Mol. Biol.* **6**: 535–542.

Fine, J.M., Gordon, T., Chen, L.C., Kinney, P., Falcone, G. and Beckett, W.S. (1997) Metal fume fever: characterization of clinical and plasma IL-6 responses in controlled human exposures to zinc oxide fume at and below the threshold limit value. *J. Occup. Environ. Med.* **39**: 722–726.

Gilmour, P.S., Brown, D.M., Lindsay, T.G., Beswick, P.H., MacNee, W. and Donaldson, K. (1996) Adverse health effects of PM_{10} particles: involvement of iron in generation of hydroxyl radical. *Occup. Environ. Med.* **53**: 817–822.

Heinrich, U., Fuhst, R., Rittinghausedn, S., Creutzenberg, O., Bellmann, B., Koch, W. and Levsen, K. (1995) Chronic inhalation exposure of Wistar rats and two different strains of mice to diesel engine exhaust, carbon black and titanium dioxide. *Inhalation Toxicol.* **7**: 533–556.

Johnston, C.J., Finkelstein, J.N., Gelein, R., Baggs, R. and Oberdörster, G. (1996) Characterization of the early pulmonary inflammatory response associated with PTFE fume exposure. *Toxicol. Appl. Pharmacol.* **140**: 154–163.

Kittelson, D.B. (1998) Engines and nanoparticles: a review. *J. Aerosol Sci.* **29**: 575–588.

Kuschner, W.G., D'Alessandro, A., Wong, H. and Blanc, P.D. (1997a) Early pulmonary cytokine responses to zinc oxide fume inhalation. *Environ. Res.* **75**: 7–11.

Kuschner, W.G., Wong, H., D'Alessandro, A., Quinlan, P. and Blanc, P.D. (1997b) Human pulmonary responses to experimental inhalation of high concentration fine and ultrafine magnesium oxide particles. *Environ. Health Perspect.* **105** (11): 1234–1237.

Kuschner, W.G., D'Alessandro, A., Wintermeyer, S.F., Wong, H., Boushey, H.A. Blanc, P.D. (1995) Pulmonary responses to purified zinc oxide fume. *J. Investigative Med.* **43**: 371–378.

Lee, K.P., Trochiomowicz, H.J. and Reinhardt, C.F. (1985) Pulmonary response of rats to titanium dioxide (TiO_2) by inhalation from two years. *Toxicol. Appl. Pharmacol.* **79**: 179–192.

Li, X.Y., Gilmour, P.S., Donaldson, K. and MacNee, W. (1996) Free radical and pro-inflammatory activity of particulate air pollution (PM_{10}) in vivo and in vitro. *Thorax* **51**: 1216–1222.

Li, X.Y., Brown, D., Smith, S., MacNee, W. and Donaldson, K. (in press) Overload studies with fine and ultrafine carbon black in rat lungs after intratracheal instillation. *Inhalation Toxicol.*

MacNee, W., Li, X.Y., Seaton, A. and Donaldson, K. (submitted) Effects of short-term low exposure to carbon black: enhanced effect of ultrafine carbon black. *Am. J. Resp. Crit. Care. Med.*

Mauderly, J.L., McCunney, R.J. and Eds (1996) Supplemental issue on Particle Overload in the rat lung and lung cancer: implications for risk assessment. *Inhalation Toxicol.* **8** (Suppl): 1–298.

Maynard, R.L. and Waller, R. (1996) Suspended particulate matter and health: new light on an old problem. *Thorax* **51**: 1174–1176.

Murphy, S.A., Berubé, K.A., Pooley, F.D. and Richards, R.J. (1998) The response of lung epithelium to well characterised fine particles. *Life Sciences* **62**: 1789–1799.

Nikula, K.J., Avila, K.J., Griffith, W.C. and Mauderly, J.L. (1997) Lung tissue responses and sites of particle retention differ between rats and cynomolgus monkeys exposed chronically to diesel exhaust and coal dust. *Fund. Appl. Toxicol.* **37**: 37–53.

Oberdörster, G. (1996) Significance of particle parameters in the evaluation of exposure dose–response relationships of inhaled particles. *Inhalation Toxicol.* **8** (Suppl): 73–89.

Oberdörster, G., Ferin, J. and Lehnert, B.E. (1994) Correlation between particle-size, in-vivo particle persistence, and lung injury. *Environ. Health Perspect.* **102**.

Oberdörster, G., Gelein, R.M., Ferin, J. and Weiss, B. (1995) Association of particulate air-pollution and acute mortality–involvement of ultrafine particles. *Inhalation Toxicol.* **7**.

Peters, A., Wichmann, H.E., Tuch, T., Heinrich, J. and Heyder, J. (1997) Respiratory effects are associated with the number of ultrafine particles. *Am. J. Respir. Crit. Care Med.* **155**: 1376–1383.

Pope, C.A., Bates, D.V. and Raizenne, M.E. (1995) Health effects of particulate air pollution: time for reassessment?. *Environm. Health Perspect.* **103**: 472–480.

Pritchard, R.J., Ghio, A.J., Lehmann, J.R., Winsett, D.W., Tepper, J.S., Park, P., Gilmour, M.I., Dreher, K.L. Costa, D.L. (1996) Oxidant generation and lung injury after particulate air pollutant exposure increase with the concentrations of associated metals. *Inhalation Toxicology* **8**: 457–477.

Rahman, I. and MacNee, W. (1998) Role of transcription factors in inflammatory lung diseases. *Thorax* **53**: 601–612.

Rahman, I., Morrison, D., Donaldson, K. and MacNee, W. (1996) Systemic oxidative stress in asthma, COPD and smokers. *Am. J. Resp. Crit. Care. Med.* **154**: 1055–1060.

Samet, J.M., Reed, W., Ghio, A.J., Devlin, R.B., Carter, J.D., Dailey, L.A., Bromberg, P.A. and Madden, M.C. (1996) Induction of prostaglandin-h synthase-2 in human airway epithelial cells exposed to residual oil fly-ash. *Toxicol. Appl. Pharmacol.* **141**: 159–168.

Samet, J.M., Stonehuerner, J., Reed, W., Devlin, R.B., Dailey, L.A., Kennedy, T.P., Bromberg, P.A. and Ghio, A.J. (1997) Disruption of protein tyrosine phosphate homeostasis in bronchial epithelial cells exposed to oil fly ash. *Am. J. Physiol–Lung Cell Mol Physiol* **16**: L426–L432.

Schlesinger, R.B., Bender, J.R., Dahl, A.R., Snipes, M.B. and Ultman, J. (1998) Deposition of inhaled toxicicants. In: *Handbook of Human Toxicology* (ed. E.J. Massaro). CRC Press, Boca Raton, FL, pp. 493–550.

Seaton, A., MacNee, W., Donaldson, K. and Godden, D. (1995) Particulate air-pollution and acute health-effects. *Lancet* **345**: 176–178.

Sioutas, C., Koutrakis, P., Godleski, J.J., Ferguson, S.T., Kim, C.S. and Burton, R.M. (1997) Fine particle concentrators for inhalation exposures–Effect of particle size and composition. *J. Aerosol Sci.* **28**: 1057–1071.

Stone, V., Shaw, J., Brown, D.M., MacNee, W., Faux, S.P. and Donaldson, K. (1998) The role of oxidative stress in the prolonged inhibitory effects of ultrafine carbon black on epithelial cells function. *Toxicol. In Vitro.* **12**: 649–659.

Stone, V., Tuinman, M., Vamvakapoulos, Faux, S.P., Borm, P., MacNee, W. and Donaldson, K. (submitted) Increased calcium influx on exposure to ultrafine carbon black: a potential mechanism for the pathogenicity of ultrafine particles.

Tran, C.L., Jones, A.D., Cullen, R.T. and Donaldson, K. (1998) Influence of particle characteristics on the clearance of low toxicity dusts from the lungs. *J. Aerosol Sci.* **29** (Suppl 1): S1269–S1270.

Veronesi, B., Oortgiesen, M., Carter, J.D. and Devlin, R.B. (1999) Particulate matter initiates inflammatory cytokine release by activation of capsaicin and acid receptors in a human bronchial epithelial cell line. *Toxicol. Appl. Pharmacol.* **154**: 106–115.

Yeh, H.C., Muggenburg, B.A. and Harkema, J.R. (1997) In vivo deposition of inhaled ultrafine particles in the respiratory tract of rhesus monkeys. *Aerosol Sci. Technol.* **27**: 465–470.

Zhang, Q., Kusaka, Y., Sato, K., Nakakuki, K., Kohyama, N. and Donaldson, K. (1998) Differences in the extent of inflammation caused by intratracheal exposure to three ultrafine metals: role of free radicals. *J. Toxicol Environ. Health* **53**: 423–438.

Zhang, Q., Kusaka, Y., Sato, K., Mo, Y., Fukuda, M. and Donaldson, K. (1998b) Toxicity of ultrafine nickel particles in lungs after intratracheal instillation. *Occup. Health* **40**: 171–176.

Zhang, Q., Kusaka, Y., Sato, K., Mo, Y. and Donaldson, K. (in preparation) Enhanced pulmonary TNF production underlies the increased inflammation caused by ultrafine nickel compared to standard nickel. *Toxicol. Appl. Pharmacol.*

9 Sources and behaviour of atmospheric particulate matter

Roy M Harrison

9.1 Introduction

Atmospheric particles exist in a very wide range of sizes and have a variety of chemical compositions (Harrison and van Grieken, 1998). As outlined in earlier chapters, airborne particles typically range from the smallest transient nuclei of around 1 nm diameter to very coarse particles of the order of 100 μm diameter. The size distribution is typically trimodal, this distribution reflecting the sources and behaviour of atmospheric particles. It is therefore pertinent to discuss it briefly here. The three modes in the size distribution are as follows:

- *Transient nuclei generally smaller than 100 nm in diameter*. These are freshly formed or emitted particles which can arise from gas-to-particle conversion processes (e.g. oxidation of sulphur dioxide to sulphuric acid), or may be recently emitted from a combustion or other high temperature source in the urban atmosphere, most commonly vehicle exhaust. These particles are of rather short atmospheric lifetime due to growth by coagulation and condensation processes. They are, however, seen in abundance close to road traffic and in plumes from combustion sources.
- *Accumulation mode*. This size mode ranges typically from around 100–200 nm at the lower extreme, up to 2.5 μm at its upper end. These particles arise from the growth of nucleation mode particles and are relatively long-lived in the atmosphere because their growth rates are very modest and the removal mechanisms which affect other particles such as dry and wet deposition are very inefficient in this size range. A

Particulate Matter, edited by R.L. Maynard and C.V. Howard

substantial proportion of urban particles are therefore in the accumulation mode (usually at least 50% by mass other than in very dry climates) and because of their atmospheric lifetime, typically of a week or more, these particles are subject to long-range atmospheric transport contributing to regional pollution problems.
- *Coarse particles: particles of greater than about 2.5 μm.* These arise primarily from mechanical attrition processes, which include the break-up of the sea surface into sea spray (marine aerosol) and resuspension of land surface dusts (e.g. soil and street dust).

Whilst these sub-divisions of particles are not entirely clear-cut because, for example, of coagulation between very fine and very coarse particles, the size ranges are sufficiently distinct that measurements of size distribution and the associated chemistry can themselves give a useful estimation of particle origins. Determination of the source apportionment of atmospheric particles is far more complex than for the majority of other pollutants because of the formation of particles within the atmosphere from gas-to-particle conversion (termed secondary particles) and the rather poorly quantified nature of some of the sources of primary particles, especially those relating to coarse particles such as resuspension of surface soils and dusts and release from construction and demolition activities.

This chapter will address the issue of the source apportionment of particles in the atmosphere of the UK. The methodologies used are generic and therefore applicable internationally. The results will vary according to national conditions, but are broadly representative of mid-latitude developed countries. For obvious reasons quite different relative source contributions and particle abundances may arise in different climatic regions and in less developed countries where pollution abatement has not advanced to the extent of the developed world. At the time of publication of the *Third Report of the Quality of Urban Air Review Group* (QUARG, 1996), knowledge of the source attribution of particles was rather basic, and from measurements within the atmosphere, it was possible to identify three main groupings of particles:

- Primary combustion particles, mostly from road traffic existing predominantly in the fine (<2.5 μm) range.
- Secondary particulate matter, predominantly ammonium sulphate and ammonium nitrate formed by gas-to-particle conversion in the atmosphere and existing mainly in the fine (<2.5 μm) size range. Because of their rather slow formation in the atmosphere, secondary particles are far more spatially uniform than primary particulate matter which has elevated concentrations close to local sources.
- Coarse particles attributed mainly to wind-blown suspension of surface soils and dusts and to sea spray (even at considerable distances inland). The subject of source apportionment of particles in the UK atmosphere was revisited in 1998 by the Atmospheric Particles Expert

Group (APEG, 1999). This chapter will summarize some of the main findings of the group as well as referring to relevant results of recent research work.

9.2 Source apportionment of airborne particles

This refers to the quantitative attribution of ground-level concentrations to the source categories responsible for them. There are three main methods available, all of which have been used in generating the results summarized here. These methods are as follows:

- *Emissions inventories*. These are derived from quantifying the levels of an activity, for example the number of vehicle miles travelled annually or the number of kilowatt hours of electricity generated in coal-fired power stations, and multiplying this quantity by an emission factor expressing the mass of particles emitted per unit of activity (e.g. grams of particle per vehicle kilometre for a given vehicle type). The inventories are extremely useful in deriving the split between emissions from similar sources, for example, between petrol and diesel cars and between light duty and heavy duty vehicles. This is because these sources have a similar spatial distribution and height above the ground and therefore behave similarly after emission. The inventories do of course include sources from a range of geographic locations and heights of emission, and whilst useful in describing the overall mass of particles produced, they do not allow estimation of contributions to ground-level concentrations directly from sources of different location or height.
- *Numerical models of dispersion and atmospheric chemistry*. Primary particles emitted all over Europe can contribute to airborne particle concentrations within UK cities, although the contribution per unit emission becomes progressively smaller with increasing distance. Therefore, in order to model the impact of primary particle emission on ground-level airborne concentrations, it is necessary to apply atmospheric dispersion models, taking account of not only local emissions, but emissions elsewhere within the UK, and indeed across western Europe. This cannot be achieved without considerable complexity and will usually entail nesting fine grid resolution models of local processes within models with coarser grid resolution describing emissions at greater distance. In order to model the production of secondary particles, the emissions of precursor gases must be included in a numerical model of atmospheric chemistry which will predict their oxidation rates. The latter requires knowledge of the emissions of a wide variety of substances which are not themselves converted into particulate matter since these compounds influence the oxidizing capacity of the atmosphere. The models can account for contributions from sources in different locations and with differing emission heights

as well as being able to predict concentrations of secondary particles which are not available from emissions inventories. Their application is however complex and in some instances imprecise.

- *Receptor modelling*. This term describes a range of techniques which use the measured concentrations in the air at ground-level to infer source contributions. Whilst many of the methods are rather crude, their dependence on measured data can often give considerable confidence in the results generated.

9.3 Primary particulate matter

Emissions inventories are powerful in describing the magnitude of primary particle emissions. In the case of some sources such as road traffic they are sophisticated, taking account of different vehicle types, traffic speeds *et cetera*, whilst for less easily quantified sources such as fugitive dust releases and construction and demolition activity, the estimates are very approximate. *Table 9.1* shows the estimated UK emissions of PM_{10} disaggregated by UNECE source category. In interpreting this table, it must be borne in mind that the sources included are in a wide variety of locations and have different emission heights. A comparison with the urban inventory for Greater London (*Table 9.2*) is instructive in the former regard in that within Greater London road traffic accounts for 77% of total estimated emissions against 25% in the National Atmospheric Emissions Inventory. This, together with the fact that road traffic emissions occur at ground-level and hence have a greater impact on ground-level concentrations per tonne emitted than emissions from a high level source such as a power station, implies a very much greater role for road traffic in generating urban ground-level pollution than would be inferred from the inventories. As mentioned above, however, the inventories are extremely useful in delineating the relative contributions of closely related source categories to ground-level concentrations. Projected future emissions of particulate matter from the road transport source category have been estimated by the National Atmospheric Emissions Inventory and are shown in *Figure 9.1*. This indicates that particulate matter emissions from road traffic peaked in 1989, since which they have reduced significantly, but are set to reduce quite dramatically over the next 10 years or so as a result of the tightening of emissions standards for new vehicles and the introduction of cleaner fuels.

As mentioned above, not all primary particulate matter in the UK atmosphere is generated locally. A simple trajectory model shows that on an annual average basis about 1.9 $\mu g\ m^{-3}$ of PM_{10} in London and 1.0 $\mu g\ m^{-3}$ of PM_{10} in Edinburgh is attributable to transport of primary particulate matter from mainland Europe. This is a modest but not negligible contribution to annual averages typically in the range of

Table 9.1. UK emissions of PM_{10} by sector (ktonnes) (APEG, 1999)

Source	1970	1975	1980	1985	1990	1991	1992	1993	1994	1995	1996	1996 as %
Combustion in energy production and transformation	78	61	60	53	60	59	59	57	49	40	38	18%
Public power	58	54	54	49	57	56	55	53	46	36	35	16%
Other industrial	20	7	6	4	3	3	3	3	3	3	3	2%
Commercial, institutional and residential combustion	235	139	109	99	55	59	53	54	44	34	36	17%
Domestic	216	127	98	89	48	51	46	47	38	28	30	14%
Other	19	11	11	10	8	8	7	7	7	6	6	3%
Industrial combustion	74	43	33	25	26	26	27	25	23	20	17	8%
Iron and steel	9	5	2	1	1	1	1	1	1	1	1	1%
Other	65	38	30	24	24	25	26	24	22	19	16	8%
Non-combustion processes	57	58	54	56	63	60	58	59	61	59	59	28%
Industrial processes	33	32	31	32	32	32	31	31	32	32	32	15%
Mining and quarrying	21	23	20	22	28	25	23	24	26	24	24	11%
Construction	3	3	3	3	4	4	4	4	4	4	4	2%
Road transport	46	50	55	58	67	66	64	61	60	56	52%	25%
Petrol exhaust	11	13	15	18	18	16	15	14	13	12	11	5%
Diesel exhaust	33	35	37	37	45	46	44	43	43	40	37	17%
Tyre wear	0	0	0	0	1	1	1	1	1	1	1	0%
Brake wear	2	2	2	3	4	4	4	4	4	4	4	2%
Other transport and machinery	5	5	5	5	5	5	5	4	4	4	4	2%
Waste treatment and disposal	44	44	46	45	38	37	34	33	26	7	6	3%
Total	538	400	362	341	314	313	299	293	267	220	213	100%

Table 9.2. Estimated emissions of PM_{10} in London (tonnes), from London Research Centre (1998)

	London
Combustion in energy production	343
Commercial and residential combustion	238
Industrial combustion	182
Industrial processes	400
Solvent use	47
Road transport	7649
Cars and taxis	1554
LGVs	632
HGVs	4871
Buses	570
Motorcycles	22
Other transport and machinery	537
Waste treatment and disposal	491
Total	9887
% Road transport–All sources	77%
% Road transport–Excluding industrial processes	83%

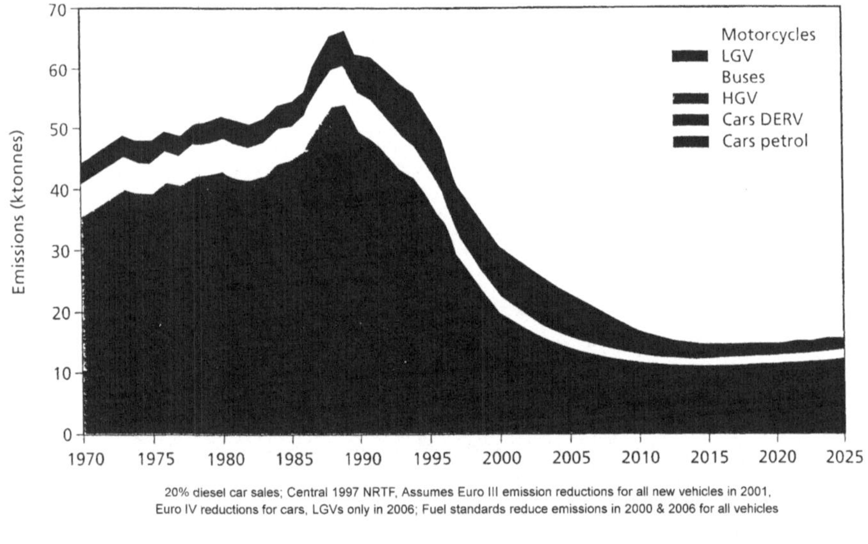

Figure 9.1. Projected UK emissions of PM_{10} from the road traffic sector (from APEG, 1999).

20–30 $\mu g\ m^{-3}$. The model, when run with a size disaggregated emissions inventory, indicates that the majority of particles transported from mainland Europe are within the $PM_{2.5}$ size range.

Analysis of data from roadside sites in comparison to nearby urban background sites indicates contributions of the road of up to 15 $\mu g\ m^{-3}$ to the annual average concentration of PM_{10} at the roadside site. These contributions are highly variable and no doubt much influenced by the precise location of the sampler with respect to the road and the prevailing wind direction. Hourly average differences between roadside and urban background concentrations for such pairs of sites can be very much greater, extending up to in excess of 50 μg m^{-3} for the London Marylebone Road site in comparison with urban background measurements at London, Bloomsbury. Comparison of both PM_{10} and $PM_{2.5}$ data from these sites shows an appreciable elevation in both fine (<2.5 μm) and coarse (2.5–10 μm) particles. Whilst it can safely be assumed that the fine particles arise from vehicle exhaust emissions, the source of the coarse particles is less obvious, but is thought to be due to resuspension of surface dusts by the turbulence generated by passing traffic (APEG, 1999). Comparison of concentrations measured at the side of major UK motorways with nearby background stations indicates that the motorways located in open country contribute around 1 μg m^{-3} for every 1000 vehicles per hour as an annual average. Interestingly, at these and some other heavily trafficked sites examination of the diurnal profiles of the pollutant concentrations shows an appreciable difference in the diurnal pattern of carbon monoxide (emitted predominantly by petrol vehicles) and oxides of nitrogen to which diesels make a far more substantial contribution. The diurnal profile of particulate matter is generally closer to that of oxides of nitrogen reflecting the major contribution of diesels.

Evaluation of coarse mode particle concentration data shows periods, especially in summer, when high wind speeds are associated with high concentrations of airborne coarse particles. This is at first sight unexpected since high wind speeds lead to dilution of locally emitted particles. It appears that the cause is the wind-driven resuspension of surface dusts which can lead to situations where coarse particle concentrations can exceed 200 μg m^{-3} hourly average and greatly exceed those of fine particles (Harrison *et al.*, 1997).

Studies conducted as part of the pilot phase of local air quality management in the UK have also revealed substantial non-traffic local sources of PM_{10} such as quarrying, local industry, and in the case of Belfast, combustion of solid fuels.

9.4 Secondary particulate matter

Secondary particles make a substantial contribution to airborne concentrations throughout the UK. They are spatially more uniform than primary particles showing rather small urban-to-rural gradients unlike vehicle-emitted particles. Their temporal variations in concentration are, however, episodic but not necessarily correlating with episodes in primary particles.

The predominant components of secondary particulate matter are as follows:

- Sulphate formed from sulphur dioxide oxidation and present in the atmosphere mainly as ammonium sulphate. Measurements in Leeds show average urban concentrations of sulphate of 5.5 μg m^{-3} in 1995–1996.

- Nitrate arising from oxidation of NO_x in the atmosphere and comprising mainly ammonium and sodium nitrates. The study in Leeds in 1995–1996 revealed an average concentration of 4.0 μg m^{-3} of nitrate.
- Secondary organic matter derived from the oxidation of volatile and semi-volatile organic matter. This has been extremely poorly characterized in the UK and within Europe as a whole, but a small dataset from Birmingham indicates a winter mean of 0.8 μg m^{-3} and a summer mean of 3.7 μg m^{-3} of PM_{10} mass. The appreciable summer-to-winter ratio is the result of more rapid photochemically induced oxidation of volatile organic matter in the summer months to form particulate matter.

Based on a network of sulphate measurements, and assuming a similar spatial distribution of ammonium sulphate and ammonium nitrate, *Figure 9.2* shows the annual average distribution of secondary sulphates and nitrates in the atmosphere of the UK. There is clear gradient from south-east to north-west reflecting the proximity of the south-east to long-range transport of secondary particles and their precursors from the European mainland and being on the downwind side of the UK and hence having more time for formation of secondary particles from precursor gases on air-mass trajectories from the west.

In order to gain insights into the episodicity of secondary inorganic particulate matter, the Meteorological Office has developed a Lagrangian trajectory model incorporating atmospheric chemistry which can simulate the production of sulphate particles at any receptor point in the UK with a time resolution of 15 min. The model is able to pinpoint specific

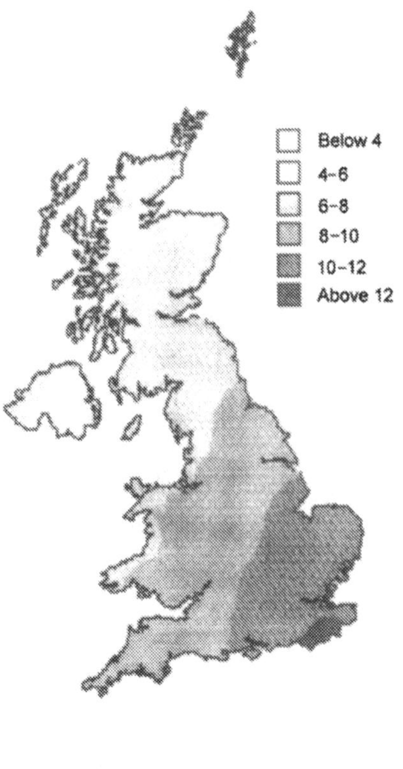

Figure 9.2. Estimated annual mean secondary inorganic PM_{10} concentration, 1996 (μg m^{-3}).

sources of precursors of airborne sulphate and can therefore apportion sulphate to sulphur dioxide emissions in mainland Europe and in the UK. In the extreme air pollution episode of March 1996 which was dominated by secondary particulate matter, the model shows that for sulphur in London, 68% arose from sulphur dioxide emitted in mainland Europe, whilst for Birmingham 55% was from mainland European sources. In the case of the more typical July 1996 episode, sulphate arose mainly from sulphur dioxide emitted in the UK with 22% in London and 13% in Birmingham arising from mainland European sources. Over an entire year, both UK and mainland sources make a substantial contribution.

It should be noted that some secondary components of airborne particulate matter, most notably ammonium nitrate and many of the secondary organic materials, are semi-volatile. Some primary components are also semi-volatile but this is of lesser importance. Some measurement instruments for PM_{10} and $PM_{2.5}$, most notably the tapered element oscillating microbalance (TEOM) samplers, pre-heat the inlet air to 50°C prior to filtration. This causes substantial loss of semi-volatile material and these instruments give lower measurements of PM_{10} and $PM_{2.5}$ than co-located samplers which do not heat the inlet air prior to filtration. TEOM instruments are used within the UK National Automatic Network and this influence is both well quantified and accounted for in standard setting.

9.5 Receptor modelling of airborne particles

As indicated above, a wide range of techniques is available for receptor modelling. In their simplest form receptor models use another pollutant as a tracer of particles from a given source category. In this context, carbon monoxide in the UK arises very largely from road traffic emissions, as does NO_x within some of the conurbations. These gases are therefore useful as a tracer of road traffic emissions. A technique has been outlined by Deacon *et al.* (1997) which uses regression analysis of particulate matter data with carbon monoxide and NO_x data to estimate the contribution of road traffic emissions to PM_{10} at urban background sites. This approach reveals a range of traffic-generated PM_{10} between 2 and 17 $\mu g\ m^{-3}$ at UK urban background sites for specific seasons of the year. These represent, on average across the country, about one-third of the total concentrations of PM_{10} as measured by the TEOM instrument. If days are selected on which especially strong correlations are seen between PM_{10} and the traffic-related gases, the fraction of PM_{10} which is traffic related is typically around 75% and higher on days with PM_{10} in excess of 50 $\mu g\ m^{-3}$. Thus, traffic emissions can contribute appreciably to episodes of high PM_{10} concentration especially in the winter months.

Particle number can also be used as a tracer for vehicle emissions. Because of the substantial contribution of secondary particulate matter to concentrations of PM_{10}, increments due to local sources such as road traffic are often difficult to discern in continuous records of particle concentration. However, vehicles are an extremely strong source of new particles, and if particulate matter is evaluated by number concentration (i.e. number of particles per cubic centimetre of air), emissions from road vehicles and from other combustion sources are seen very clearly above the background of accumulation mode particles which contribute far less to particle number concentration. An example of this is seen in *Figure 9.3* which shows the average diurnal variation of particle number count together with PM_{10} and traffic-generated gases on the Marylebone Road in London. PM_{10} concentrations are seen to be a strong indicator of emissions from road traffic, in this case primarily diesel vehicles since they correlate more closely with NO_x than with carbon monoxide. *Figure 9.4* shows for different vehicle categories the percentage of vehicle numbers at a site in central Oxford represented by the specific category, the contribution to PM_{10} mass emissions and to particle number, each as a percentage attributable to this vehicle category. Thus, whilst gasoline (petrol) vehicles dominate by number the overall traffic and make the largest contribution to PM_{10} mass, their contribution to particle number is tiny compared to that of the diesel-fuelled vehicles. Buses, which represent only 3% of vehicle numbers, contribute 54% of particle numbers at this site.

A valuable technique developed by Stedman (1998) allows discrimination on a daily basis between combustion-generated particles (mainly road traffic generated), secondary particles and coarse particles. The method uses measurements of black smoke concentration to evaluate the

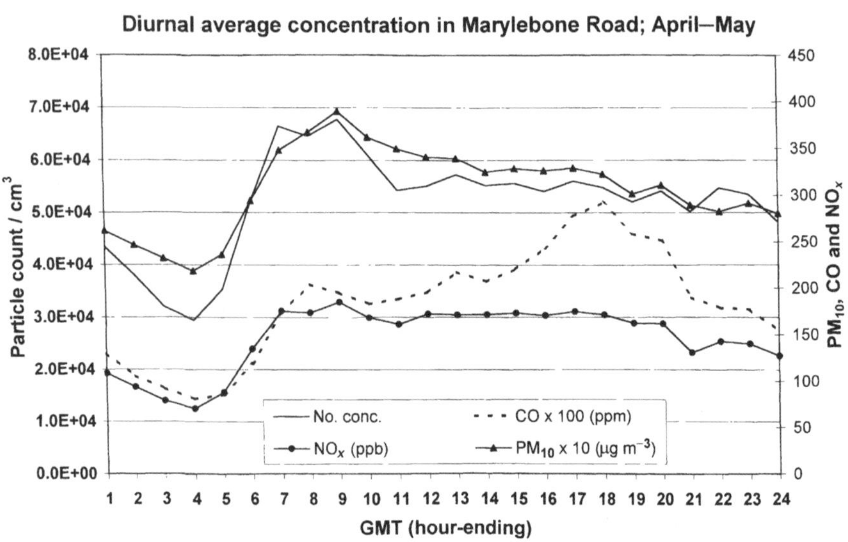

Figure 9.3. Average diurnal variation of PM_{10} mass and particle number count together with NO_x at Marylebone Road London.

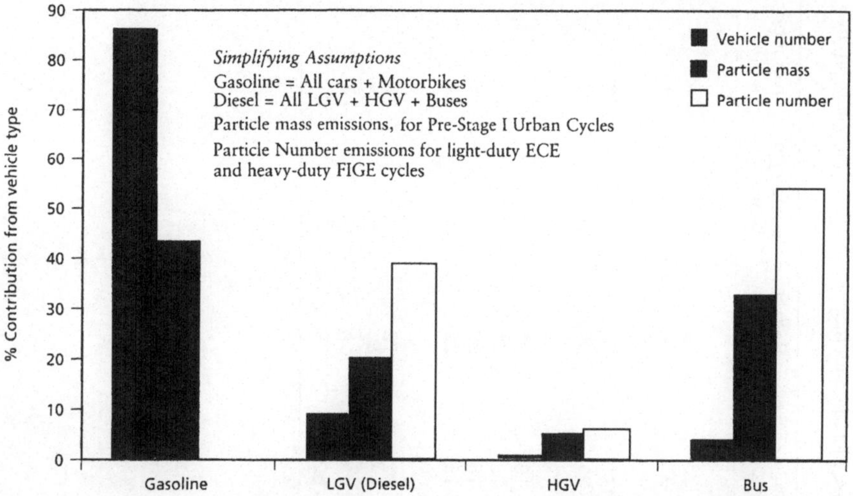

Figure 9.4. Assignment of particle mass and particle number as measured at Carfax, Oxford to different vehicle categories (measurements by AEA Technology reported in APEG, 1999).

concentration of combustion particles, and of sulphate as an indicator of secondary particles, and coarse particles are determined by difference from the total PM_{10} mass. This technique gives a day-to-day breakdown in the contributions of the various source categories and clearly demonstrates episodes of high PM_{10} concentration in which combustion particles and sulphate particles are separately dominant as well as episodes where both make an appreciable contribution, and in some instances where coarse particles are also substantial contributors. *Table 9.3* gives a breakdown of the number of days at selected sites at which a concentration of 50 μg m^{-3}

Table 9.3. Source apportionment of days with PM_{10}>50 μg m^{-3} at six monitoring sites (APEG, 1999)

Number of days	Strategy	Total	Primary	Secondary	'Other'	p+s	p+o	s+o	p+s+o
1995									
London Bloomsbury	46	26	3	2	18			3	
Birmingham centre	24	11	3	1	6		1		
Bristol centre	27	12	3	4	1		2	2	
Manchester Piccadilly		–	–	–	–	–	–	–	–
Newcastle centre	20	9	1	4	2			1	1
Belfast centre	25	13	7		4		1	1	
1996									
London Bloomsbury	56	37	5	21	4			2	5
Birmingham centre	36	20		12	4		2	1	
Bristol centre	37	18	1	7	5	2		1	
Manchester Piccadilly	42	14	1	14	4				2
Newcastle centre	27	14		13				1	
Belfast centre	31	18	8	7	3				

The number of days with a maximum 24 hour running greater than or equal to 50 μg/cm^3 is shown in column labelled 'strategy'. 'Total' refers to the number of days with a mean daily concentration greater than or equal to 50 μg/cm^3.

Table 9.4. Estimated annual means and 99th percentiles of PM_{10} concentrations at urban background sites, including predictions for 2005 and 2010 ($\mu g\ m^{-3}$) (APEG, 1999)

	London, Bloomsbury	Birmingham, Centre	Bristol, Centre	Manchester, Piccadilly	Newcastle, Centre	Belfast, Centre	Edinburgh, Centre	Liverpool Centre
Annual mean								
1996	30	25	26	26	23	24	19	25
1995	28	23	24	–	22	24	20	27
NAQS 99th percentile								
1996	100	94	87	85	100	94	62	86
1995	80	77	70	–	71	95	61	72
Annual mean								
2005 estimate based on 1996	25	20	22	22	19	21	17	21
2005 estimate based on 1995	23	19	19	–	18	21	18	22
2010 estimate based on 1996	23	18	20	20	18	20	16	19
2010 estimate based on 1995	21	17	18	–	16	20	17	21
NAQS 99th percentile								
2005 estimate based on 1996	82	76	70	71	82	76	55	70
2005 estimate based on 1995	69	65	57	–	57	82	57	63
2010 estimate based on 1996	76	69	64	65	72	70	53	64
2010 estimate based on 1995	65	58	52	–	52	79	55	61

was exceeded and the number of days on which the predominant contribution was from either primary combustion particles, secondary particles, coarse particles or a combination of two or three of these.

9.6 The future

The disaggregation of ambient concentrations into contributions from various source categories allows a projection of future changes in concentration from predictions of the reductions effected in particles from the various source categories by policy measures. *Table 9.4* shows the best estimate of the APEG Group of the future concentrations of PM_{10} at UK urban background locations.

Currently, UK and European regulatory action is focused predominantly on PM_{10} as a metric for airborne particulate matter. PM_{10} contains a significant fraction of coarse particles (typically about one-third at urban background sites) which are probably of far less health consequence than the fine particles, and which practicable control methods give little prospect of limiting substantially. Additionally, research work such as that of Seaton, Donaldson and co-workers (Seaton *et al.*, 1995) indicates that it may be the ultrafine particle fraction and therefore the particle number, or conceivably the surface area of particles, which is a stronger determinant of the health impact than the PM_{10} mass. Until clearer information is available on the precise determinant, be it chemical or physical, of the adverse health impact of particulate matter, it is difficult to redesign measurement and control strategies to meet the objective of safeguarding the health of the population to be more effective than current actions based on PM_{10} as the metric of airborne particles.

Acknowledgement

The content of this article is based heavily upon the report of the Airborne Particles Expert Group (APEG, 1999) for which the author acted as Chairman. Thanks are due to the members of the Expert Group for their valued contributions.

References

APEG (1999) *Source Apportionment of Airborne Particulate Matter in the United Kingdom.* The First Report of the Airborne Particles Expert Group, Department of Environment, Transport and the Regions, London.

Deacon, A.R., Derwent, R.G., Harrison, R.M., Middleton, D.R. and Moorcroft, S. (1997) Analysis and interpretation of measurements of suspended particulate matter at urban background sites in the United Kingdom. *Sci. Total Environ.* **203**: 17–36.

Harrison, R.M. and van Grieken, R. (eds) (1998) *Atmospheric Particles.* John Wiley and Sons, Chichester, 610 pp.

Harrison, R.M., Deacon, A.R. and Jones, M.R. (1997) Sources and processes affecting concentrations of PM_{10} and $PM_{2.5}$ particulate matter in Birmingham (UK). *Atmos. Environ.* **31**: 4103–4117.

London Research Centre (1998) *London Atmospheric Emissions Inventory*, ISBN 1-85261-267-3.

QUARG (1996) *Airborne Particulate Matter in the United Kingdom.* The Third Report of the Quality of Urban Review Group, Birmingham.

Seaton, A., MacNee, W., Donaldson, K. and Godden, D. (1995) Particulate air pollution and acute health effects. *Lancet* **345**: 176–178.

Stedman, J.R. (1998) The secondary particle contribution to elevated PM_{10} concentrations in the UK. *Clean Air* **28**: 87–93.

10 UK research programme on the characterization of vehicle particulate emissions

Michael J Davies

10.1 Introduction

Emissions of particulate matter are of increasing concern due to their potential acute health effects. These are expected to be dependent upon size and composition as has been discussed by other authors in this book. Although road vehicles have been highlighted as a major source of particulate matter, information on the characteristics of these emissions is very limited. Accordingly, the former Department of Transport, the former Department of the Environment and the Society of Motor Manufacturers and Traders, initiated a jointly funded research programme to obtain characteristic data for exhaust emissions and investigate robust methods for the generation of such data. This chapter summarizes the findings of this work; the full report covering progress to date has been published by ETSU (Moon and Donald, 1997). The programme has comprised three phases:

- Phase 1 – identification of existing knowledge and data on the measurement of vehicular particulate emissions through a review of the open literature.
- Phase 2 – definition of measurement techniques and instrumentation to be used in initial testing through reviews conducted by groups of experts.
- Phase 3 – generation of an initial database of vehicular exhaust particulate emission characteristics, evaluation of a range of candidate particulate sizing instruments and an assessment of the current

Particulate Matter, edited by R.L. Maynard and C.V. Howard

particulate dilution tunnel as a tool for exhaust particle concentration and size distribution analysis.

Phase 3 was formulated on the basis of the available experience and guidance given by the UK Department of Health. It was recommended that the initial data of vehicular exhaust particulate characteristics should focus on the number count in the size range 20 nm–10 μm and chemical analysis for acidity, anions, metals and organics. It was decided that the existing dilution tunnel would be used with a Scanning Mobility Particle Sizer (SMPS) and Quartz Crystal Microbalance (QCM) for particulate concentration and size distribution analysis. In parallel, an evaluation of the efficiency of the existing, legislative dilution tunnel and an investigation of a range of instruments for particle sizing would be undertaken. All three phases were managed by ETSU, with the vehicle/engine characterization being undertaken jointly by AEA Technology and Perkins Technology, and the tunnel and sizing instrument evaluations being undertaken by AEA Technology.

10.2 Road vehicle exhaust particulates

The chemistry and physics of the exhaust stream from an internal combustion engine, such as the current petrol or diesel engines, is extremely complex. Under ideal conditions the fuel and air would combust cleanly and completely to yield nitrogen (N_2), water (H_2O) and carbon dioxide (CO_2). However, in reality an extremely wide range of other species are formed and inter-react, many of which are known pollutants, these include:

- organic compounds [unburnt and partially burnt fuel and oil hydrocarbons (HC)]
- carbon monoxide (CO);
- nitrogen oxides (mainly 'thermal NO_x');
- particulates (PM);
- inorganic compounds (mainly fuel and oil derived, e.g. sulphur oxides and hydrogen sulphide).

Particulates are recognized as extremely complex, their chemistry and structure being dependent on many factors including the fuel, the actual combustion process (spark ignited, compression ignition, etc.), conditions within the cylinder prior to, during and post combustion and processes such as agglomeration, condensation, gas phase reactions and heterogeneous catalytic reactions.

10.3 Initial vehicle/engine testing

10.3.1 Test methodology

This research programme was based upon the available information obtained from the open literature and guidance from groups of experts drawn from the automotive industry (vehicle/engine manufacturers and consultants), automotive component suppliers (fuel injection and catalyst manufacturers), particulate sizing instrument manufacturers, the petroleum industry and academia (including aerosol scientists). To maintain consistency with measurement techniques for 'regulated' pollutants (gaseous hydrocarbons–HC, carbon monoxide–CO, nitrogen oxides–NO_x and for diesel only particulates–PM), samples were obtained from a dilution tunnel which complies to the relevant EC Directive for the light-duty vehicle or heavy-duty engine under test (EC Directive 70/220 and 88/77, respectively). To ensure reproducibility vehicle/engine and laboratory pre-conditioning was performed according to established best practice as developed for the EPEFE programme (European Programme on Emissions, Fuels and Engine Technologies). Standard legislatively prescribed emissions were measured for each cycle driven.

Particulate size analysis was determined using a SMPS (TSI, Minneapolis, MN) and a QCM (California Instruments, Monterrey, CA). As the scan time for the SMPS was of the order of 1 minute, initial measurements were made at steady-state (engine speed and load) conditions selected from the respective light- or heavy-duty drive cycles. From these steady-state scans three size classifications were selected to be monitored during sequential drive cycles for each vehicle/engine.

Particulate composition analysis was performed on samples collected using the standard filter paper method, as defined in the respective emissions testing legislation. Determinations were made using the techniques defined by the expert groups (Moon and Donald, 1997) (*Table 10.1*).

The Programme Steering Committee selected vehicles/engines for the initial phase (*Table 10.2*). The selection, constituting a range of types, was made to cover light-duty and heavy-duty applications, diesel and gasoline (leaded and unleaded) fuels, current and old technology and the effect of a particulate trap. This matrix was selected to provide a wide variety of types so as to develop and evaluate test methods. As only one vehicle/engine of each type was tested the results cannot be taken as representative of the type.

10.3.2 Evaluation of test methods

SMPS size and concentration measurements. Data generated from all of the vehicle and engine tests using the SMPS were considered to be robust, accurate and highly repeatable as shown by *Figures 10.1* and *10.2*.

Table 10.1.

Acidity	Wet chemical method using a modified Gran technique
Metals (especially known sensitizers)	Inductively Coupled Plasma Mass Spectroscopy (ICP-MS) and Inductively Coupled Plasma Atomic Emission Spectroscopy (ICP-AES)
Anions (such as sulphate and nitrite)	Ion chromatography (IC) -Institute of Petroleum (IP) proposed method (BT/10)
Organics	Polyaromatic Hydrocarbons (PAH–US EPA priority list of 16) by Gas Chromatography–Mass Spectrometry (GC-MS) Soluble Organic Fraction (SOF)–IP Method PM-CA

Table 10.2.

Category	Identification[a]	Description
Light-duty	1	IDI diesel, turbo-charged, non-catalyst, mid-1990s technology car
	2	IDI diesel, non-turbo, non-catalyst, 1980s technology car
	3	DI diesel, turbo-charged, oxidation catalyst, exhaust gas recirculation, mid-1990s technology van
	4	Leaded petrol, non-turbo, non-catalyst, 1980s technology car
	5	Unleaded petrol, non-turbo, three-way catalyst, mid-1990s technology car
	6	Vehicle 2 with particulate trap
Heavy-duty	7	Diesel, early-1990s technology engine
	8	Diesel, mid-1990s technology engine
	9	Engine 7 with particulate trap

[a]The numbers identify vehicles in the following figures.
IDI, Indirect injection; DI, Direct injection.

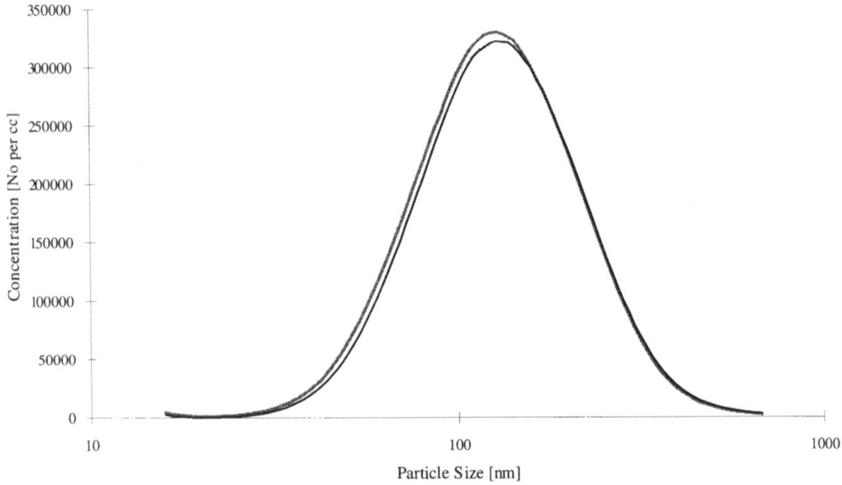

Figure 10.1. Example of repeatability of scans, SMPS particle flux at fixed road speed.

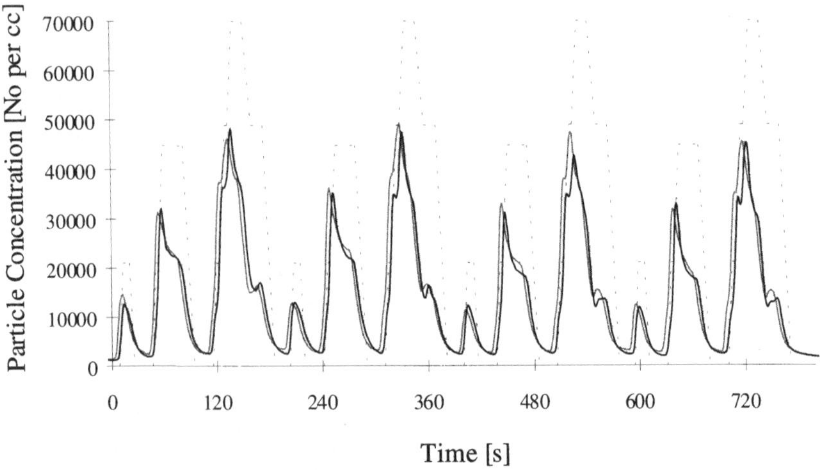

Figure 10.2. Example of repeatability of cycles, SMPS particle flux against the ECE drive cycle.

Owing to the lack of traceable absolute particle concentration standards, no calibration of the dilution tunnels, sampling probes or instruments could be made. However, using the SMPS steady-state and drive cycle data, estimates of the total number of particles emitted were derived for each vehicle tested, and assuming volume conversions and density these were compared to the regulated mass emissions. This showed that the concentrations measured using the SMPS were of the correct order of magnitude, which was considered good in view of the assumptions made. A summary of a parallel more detailed study of this instrument is reported in Section 10.4.

QCM size and concentration measurements. Data generated using the QCM were found to be far less robust. Repeatability in size measurements from day-to-day was found to be only 25–40%, depending on vehicle type. Comparing the QCM to regulated filter mass measurements showed a poor correlation. Analysis of the QCM results showed that the results obtained were actually related to the QCM efficiency, with the latter reducing significantly at higher concentrations. It was also found that as the mass of particulate collected increased the efficiency of the QCM reduced. Further investigations into the problems observed with this device were undertaken in a parallel study, the results of this are also summarized in Section 10.4.

Particulate composition measurements. The use of the modified 'Granplot' technique for the determination of particulate acidity gave only a moderately satisfactory result for the older technology heavy-duty engine (Engine No 7). Too little particulate material was collected from the other two vehicles tested [old IDI diesel vehicle (No 1) and the unleaded petrol vehicle (No 5)] to obtain a positive result. Consideration of this method, which is specific for strong acids, leads to the conclusion that particle acidity (probably largely resulting from the sulphur in the diesel fuel), can be neutralized through reactions in the exhaust, on the filter paper or during the extraction process. This could explain the low levels of acidity measured and the poor repeatability achieved.

The techniques selected for the determination of trace metals in the particulates (ICP-MS and ICP-AES) were both considered reliable for repeatable measurements. However, for many of the analytes targeted the levels measured were very similar to the 'blank' filters, alternative filter materials could reduce this problem.

Anions were measured using IC according to the IP method BT/94. This method was extended from that validated by IP to include the whole range of anions of interest. To minimize background effects the filter papers used were pre-extracted prior to collection of the vehicle/engine particulates. From the limited tests conducted, this technique was considered reliable for the repeatable determination of trace anions.

Analysis of PAHs (as identified in the US EPA list of '16 priority pollutant PAHs') was conducted by GC-MS. This technique is well proven and gave good repeatability from triplicate analyses. Having conducted blank analyses, it was found that the solvent contributed the majority of 'background' and that toluene was favoured for extraction over dichloromethane.

Determination of organic content (SOF) using IP method PM-CA/96 was found to be limited in its accuracy and repeatability due to the friable nature of the filter paper during handling. However, the repeatability for triplicate measurements was found to be fair. As a consequence, the expert group recommended that future work should determine the volatile organic fraction (VOF) instead of SOF.

For many of these compositional analyses the principal problem was found to be the collection of sufficient particulate mass to achieve an accurate measurement above limits of detection and background levels. Collection of greater amounts of particulate by running multiple cycles on a single filter paper has obvious problems associated with condensation and chemical reactions.

10.3.3 Initial results from vehicle/engine testing

As these measurements were made using a dilution tunnel connected to the exhaust pipe of the engine/vehicle under test, they cannot be taken as fully equivalent to 'real-life'. The transfer pipe and restricted dilution ratio within the tunnel are expected to increase particulate agglomeration and condensation of semi-volatiles when compared with the rapid dilution expected at the tailpipe of a road vehicle.

SMPS measurements. As described in Section 10.3.1, initially scans were taken at fixed road speeds to define the distributions of particle size expected during the legislative drive cycles. Examples of the data obtained are shown in *Figure 10.3*.

As can be seen there are significant differences between the different vehicles. The modern, unleaded petrol-fuelled vehicle (No 5) emitted particulates at a concentration two orders of magnitude lower than the other vehicles. The effect of fitting the trap to the old IDI diesel vehicle was to lower the concentration of particulate emissions by up to one order of magnitude. This result obtained with the use of a trap should not be taken as typical as the trap system used was a very early prototype and the precise condition of the trap will obviously be critical. The effect of road speed on particulate emissions is shown in *Figure 10.4*.

Again the significant effect of fitting a particulate trap can be seen (Engine 9; *Figure 10.5*). Owing to the fact that the trap fitted was very simple and no consideration of regeneration to prevent excessive backpressure on the engine had been made, this result should not be considered typical. Again the effect of engine load is significant as shown by *Figure 10.6*.

These scans were then used to select three size bands to monitor during successive drive cycle tests, *Figure 10.2* being typical of the results obtained.

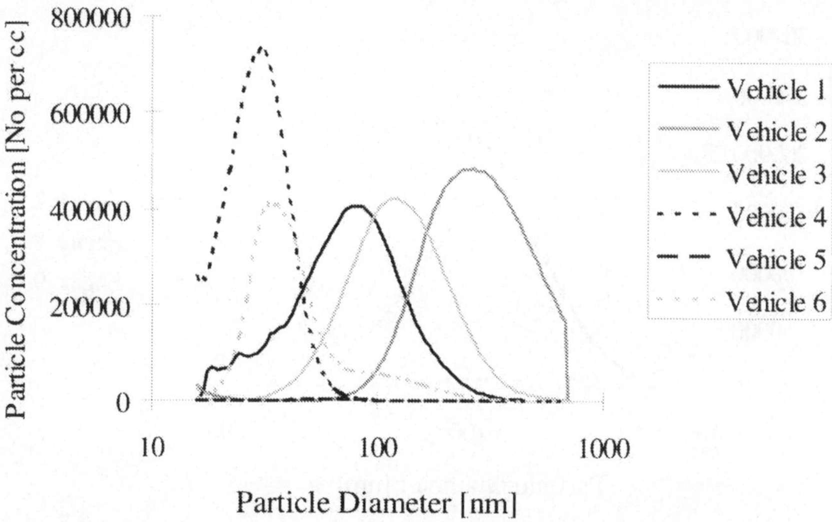

Figure 10.3. Steady-state SMPS, light-duty vehicles, 100kph

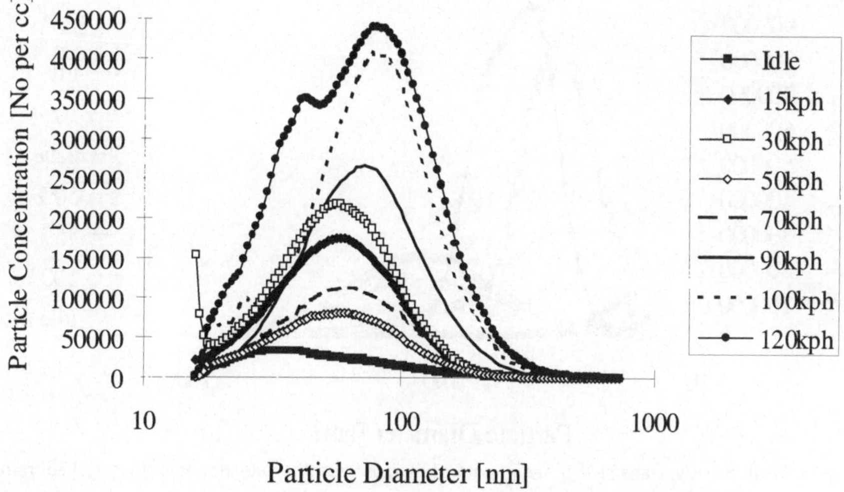

Figure 10.4. Steady-state SMPS, IDI turbo-charged diesel, variation with road speed.

10.4 Sizing instrument investigation

In parallel to the characterization of particulates emitted by road vehicles and engines, it was felt necessary, due to the lack of experience in performing particulate measurements from these sources, to undertake an investigation of instruments for particulate sizing.

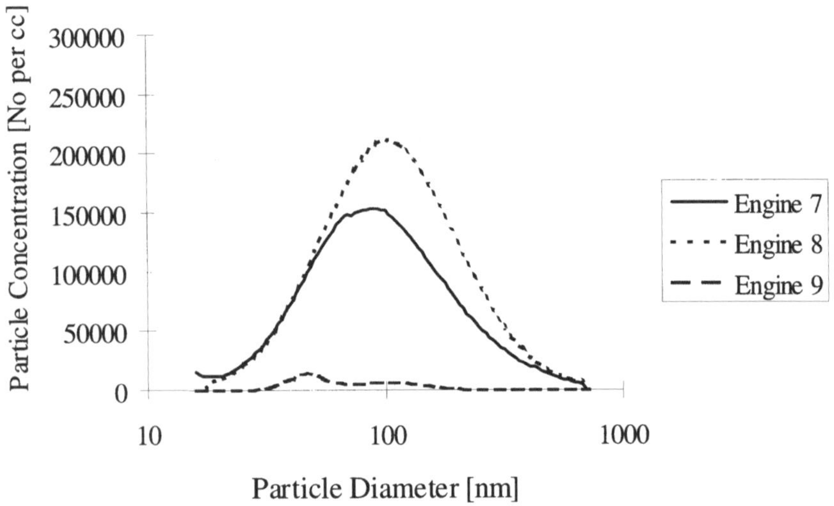

Figure 10.5. Steady-state SMPS, heavy-duty engines, 100% load @ 190rpm

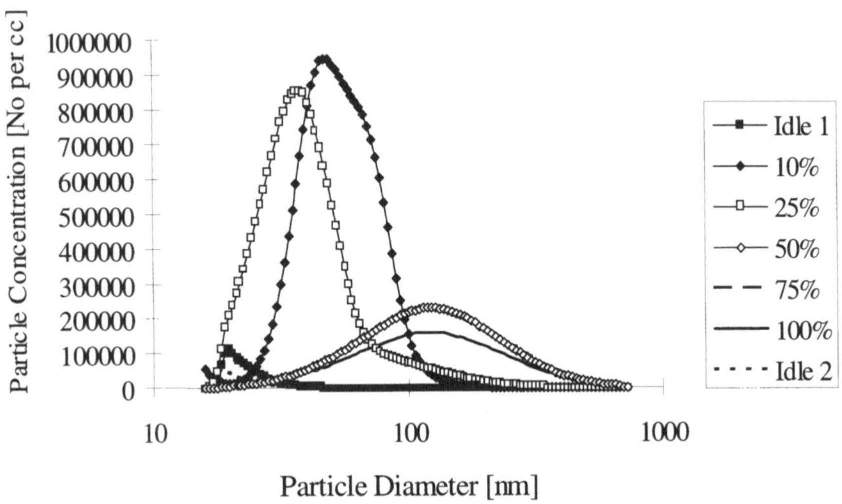

Figure 10.6. Steady-state SMPS, new heavy-duty diesel engine, variation with load at 1260rpm

10.4.1 Scanning mobility particle sizer

This instrument measures particle number distributions by electrical mobility and optical counting. High-resolution measurements (~128 channels) are provided from 13 to 750 nm, or 7 to 300 nm, or 5 to 200 nm. At full resolution a complete scan takes approximately 2.5 min although faster scans can be obtained at lower resolution, or real-time measurements can be made at a fixed size selection. Calibration was found to be excellent, although the concentration calibration was found to be systematically low by a factor of 2. The repeatability was also excellent.

10.4.2 Quartz crystal microbalance

The QCM provides measurements of particle mass distributions using 10 stages between 50 and 25 000 nm. However, in these tests, using diesel vehicle particulates, in comparison to the regulated filter paper the QCM was found to under-report by about 90%. The repeatability was found to be significantly worse than the other instruments tested. Initially it was thought that this instrument was being affected by charge on the particulates. However, even after charge equilibration, this instrument still under-performed in the measurement of diesel particulates. The QCM also reported mass size distributions that were smaller than the volume based distributions reported by both the ELPI and the SMPS. It was concluded that these issues were indicative of poor coupling to the measurement stages and high inter-stage wall losses. Therefore, the QCM was assessed as being unsuitable for the determination of vehicle exhaust particle size distributions.

10.4.3 Electrical low-pressure impactor (ELPI)

This instrument can make particulate size measurements in two modes.

Mass mode. In this mode the instrument acts as a low-pressure impactor and real-time measurements are not possible. Twelve stages are used to provide size distribution measurements between 32 and 10 000 nm. Calibration of the ELPI was considered good and during steady-state and drive cycle tests the results obtained agreed well with other instruments. However, cumulative weighing errors resulted in levels of repeatability lower than with the regulated filter paper. With a revision to the calibration of the cut-off of the lower stages good agreement was achieved with the Anderson Impactor and limited agreement with SMPS volume data.

Number mode. In this mode particles are classified by impaction and counted by charge measurement on each stage, the exhaust stream having been pre-charged. Calibration with mono-dispersed aerosols showed that the mass-mode calibration was not valid in number mode. The distributions reported by the ELPI were centred at larger sizes than by the SMPS. Calibration work by Minnesota University has generated modified cut-off values for each stage, which on initial inspection improved the ELPI-SMPS correlation. Calibration of concentration is still awaited. The repeatability of the ELPI proved to be good. It was concluded that if the calibration of this instrument was completed then the ELPI would be a reliable tool for measuring number size distributions in real time.

10.4.4 Tapered element oscillating microbalance (TEOM)

Rather than measuring particle size, the TEOM measures mass concentration in real-time by collection of the particulates on a vibrating filter and correlation of the induced shift in vibration frequency with the mass collected. Under steady-state tests the TEOM performed and correlated well to the Anderson Impactor and ELPI (mass mode). Over the drive cycle the mass determined with the TEOM differed from the regulated filter paper. It was also found necessary to smooth the real-time output to eliminate large fluctuations, this obviously results in a loss of transient detail.

10.4.5 Anderson impactor (Mark III)

The Anderson impactor is a cascade impactor that can be used to measure particle size distributions in seven stages from 400 to 6500 nm. Filter papers are used as the substrate on each stage with a final paper used to collect particulates smaller than 400 nm. For both the steady-state and drive cycle tests good correlation was found with both the ELPI and the filter paper, and without the need for greased stages (as with the ELPI in mass mode) this may provide a method for size classified compositional analysis.

10.5 Particulate dilution tunnel evaluation

EC Directives (70/220 and 88/77) describe the regulated methods for particulate emission testing. In summary the exhaust emissions are sampled by CVS where the exhaust is continuously diluted with air under controlled conditions to a fixed total volumetric flow. The air flow and total flow are measured so that the dilution ratio is known as a function of time over the respective drive cycle. A constant flow of this diluted exhaust gas is then sampled and passed through a pair of filter papers in series. To ensure that accurate, reproducible measurements are possible, the exact design of the transfer line from the vehicle/engine exhaust, the dilution tunnel, the sampling probe and system are defined.

Investigation of a replicate of a light-duty dilution tunnel system using particulate penetration techniques showed that transport of particulates from the vehicle tailpipe to the sampling plane occurs with an efficiency of more than 80% for particulates of less than 3 μm (aerodynamic diameter). The losses observed were thought to be primarily caused by the following design elements:

- The rough-walled flexible tube used to connect the vehicle/engine exhaust pipe to the dilution tunnel. The losses will increase as the length of the tube increases, or due to bends in the pipe, or if the pipe is not lagged against heat loss.
- The 'Chinese Hat' arrangement used to stop large particulates that have detached from the transfer pipe or dilution tunnel wall from entering the sampling system. This will result in poor penetration of particulates of aerodynamic diameter greater than 3 μm.

- The sampling system is not operated isokinetically and this would result in biasing of the measurements made.
- Calculations indicate that the flow rates specified would result in an unstable condition between laminar and turbulent flow regimes. This would result in small changes in flow rate causing large changes in penetration characteristics.

10.6 Particulate traps

As was shown in Section 10.3, incorporation of traps into the exhaust of a vehicle has a dramatic effect upon the emissions of particulates. The types of trap tested during these initial characterization tests represented examples of the wide range of designs that have been evaluated over a number of years by the automotive industry. These traps contain a filtration medium which itself poses a significant problem to the operation of the internal combustion engine i.e. backpressure. Excessive exhaust backpressure results in increased fuel consumption and potential engine durability problems. This is further exacerbated by the build-up of particulates within the trap, which causes increased flow restriction. To minimize this build-up of backpressure requires regeneration or cleaning of the 'filter cake' by combustion of the deposited particles. However, these carbonaceous particulates require temperatures of greater than 550°C and oxidizing conditions to burn.

A wide range of trapping media have been evaluated and developed for these applications. The main design requirements are low backpressure and low rise in flow restriction as a function of filter load, high filtration efficiency, high mechanical and thermal durability and small volume. The following list indicates some of the media that have been reported in the open literature:

- ceramic foams;
- metal wire pads;
- ceramic fibres supported on porous 'candles';
- sintered metal particle monoliths;
- ceramic wall-flow monoliths (shown in *Figure 10.7*) as used in the initial vehicle/engine tests. The flow is forced through the porous ceramic walls of the honeycomb monolith. Channel sizes of the order of 1–2 mm square and 150 mm long.

Diesel engine exhausts, although lean in terms of stoichiometry, rarely reach 550°C except under high speed and load conditions. Gasoline exhausts regularly achieve the required temperature for particulate oxidation but operate at stoichiometry. As a consequence a wide range of trap regeneration methods have been evaluated and developed with the target of achieving sufficiently frequent particulate burning to minimize any backpressure cycling. Heat addition techniques such as electrical

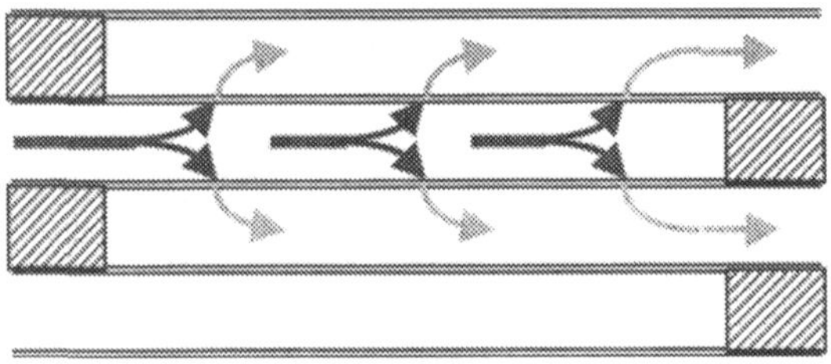

Figure 10.7. Diagrammatic representation of a ceramic wall-flow monolith.

heating, microwaves, fuel burners, and engine intake/exhaust throttling have all being developed but they result in significant fuel consumption penalties. Fuel additives, based on copper or cerium for example, which result in intimately dispersed catalytic species, all require complex dosing systems which in turn must be maintained. Catalytic coatings applied to the surface of a trap can be poisoned by sulphur species from the fuel or can cause unacceptable emission of sulphates which increase particulate mass and acidity. So-called 'continuously regenerable traps' have been developed and are under test for heavy duty application. These rely on oxidation catalysts upstream of the trap which convert nitric oxide (NO) to the strongly oxidizing agent nitrogen dioxide (NO_2). The NO_2 is able to oxidize the trapped carbonaceous particulates at significantly lower temperatures. Thereby a much wider range of engine speeds and loads result in trap regeneration. Again, unless the engine is operated on very low sulphur diesel unacceptable sulphate emissions are produced.

10.7 Conclusions

The measurement of particle size distribution and number concentration determination using the SMPS was found to be repeatable and robust for the wide range of vehicle and engine types selected. The use of the QCM however, was found to be less than satisfactory for this application. Compositional analysis for all of the analytes requested was found to be possible, however due to the low particle mass emissions from many of the engine types requires collection from large exhaust flows or over multiple cycles.

Investigation of the performance of a range of particle measurement instruments concluded that the SMPS and ELPI were suitable for particulate size distribution analysis and number-based concentration determination. The current dilution tunnel and CVS system were shown to only require small modifications to the design and operating conditions.

The initial tests suggested that exhaust particulate traps are capable of significantly reducing emissions. However, these devices need significant development to achieve the required levels of robustness, durability and efficiency whilst causing absolute minimal impact to fuel consumption and driveability under all driving conditions.

Reference

Moon, D.P. and Donald, J.R. (1997) UK Research Programme on the Characterisation of Vehicle Particulate Emissions, ETSU Report Number ETSU-R98, AEA Technology plc., Culham Laboratory, Didcot, Oxfordshire.

11 Air quality strategies with respect to particles

Martin Williams

11.1 Introduction

Before addressing the developing science surrounding particles and its implications for policies, it is worth discussing some general issues of science and policy and how the two interact. This is especially appropriate in the case of particles because, while some aspects of the issues are now common ground in scientific terms, there are still areas where our knowledge is lacking and more research needs to be carried out, before new priorities for policy can be established, and the benefits of these priorities assessed.

The major problem in marrying policy and the science which informs it is that the timescales of the two never match. This is true almost by definition, since if there were sufficient science in place, then the problem of characterizing the scientific essentials of an issue is solved and policy formulation is then determined by consideration of other issues such as the social, economic and political aspects of the problem. Unfortunately, life is generally not this simple, and one often finds that there is insufficient scientific information compared with what ideally would be required. Indeed the task of informing policy development in these situations of incomplete information, is a major challenge for those involved in the interface in government between science and policy.

There will be occasions where the need to develop or implement policy drives scientific advances, and the classic examples are to be found in the defence area, where the policy imperatives of national defence or war often force scientific and technical advances. The examples are too well known and numerous to recount here.

However, more commonly, particularly in the environmental area, an issue is brought to widespread attention either by scientific publication or

Particulate Matter, edited by R.L. Maynard and C.V. Howard

by events in the real world, or by combinations of the two. As an example of scientific discovery leading the policy process, one can cite the elucidation, in laboratory studies, in 1974 of the ozone destruction reactions of CFCs and the imputed implications for the stratospheric ozone layer. Impetus for policy development however did not gain full momentum until the discovery in 1985 of the ozone depletion in the Antarctic stratosphere. An example, closer to the concerns of this chapter, of events driving scientific research, is the well-known London smog of 1952, an undeniably major event in public health terms, but one which immediately posed searching questions to the atmospheric science and medicine of the day. Whatever the initiating process, the position will be reached in the lifetime of a policy issue, generally sooner rather than later, where scientific input is required. It is often then the case that roles may be reversed and there may be a tendency for pressure on policy to run ahead of the science, with the cry that 'something must be done'. It is in this phase that difficult challenges arise in identifying acceptable courses of action on what is often a lack of complete or definitive scientific information. In such situations it is important not to let the science and the policy response diverge unacceptably from one another. On the one hand, if the science is developing and evolving rapidly, it may be necessary to update the science and the policy response continually so that the problem does not get out of control. On the other hand, in the case where the timescales are longer, it may be necessary to design the policy response to reflect adequately the scientific uncertainties.

In such situations consideration of the Precautionary Principle arises. Amongst other things, this states that uncertainty should not be used as an excuse for postponing cost-effective action, and we are thence led to the concept of 'no regrets' policies, which for example characterized much of the climate change debate in the early 1990s. Where there are emerging scientific reasons for going further, then it is the business of those concerned with the scientific input to policy to identify and agree the scientific common ground on which the policy debate can move forward, in the sense of identifying, and quantifying where possible, the hazard and the risk assessment, before moving to the risk management phase of the process.

Identifying the common scientific ground is not a straightforward task and can be approached in a number of ways. One method which the Department of Environment, Transport and the Regions (DETR, formerly Department of the Environment, DoE) employs, is to convene expert groups to review the science on a particular subject of policy interest. These have proven very successful in the past and have produced reports which have been widely acclaimed for their coverage of issues such as acid rain (through the Review Group on Acid Rain, RGAR), ground level ozone (PORG, the Photochemical Oxidants Review Group) and the Quality of Urban Air Review Group (QUARG). DETR in March 1998 set up the Airborne Particles Expert Group (APEG) to review the rapidly evolving science on atmospheric particles and to advise on source apportionment, that is, to attempt to quantify the various source contributions to measured levels of particles in the UK. The draft APEG report was distributed widely in June 1998 for comment and peer review, in line with the recommendations in the publication on the use of science in policy (Department of Trade and Industry, 1997) for establishing 'sound science'.

The remainder of this chapter will attempt, where possible, to identify areas of common ground in scientific terms, and will highlight the challenges to the policy maker presented by the rapidly developing science of atmospheric particulate matter.

11.2 Air quality policy and health effects

In an ideal world, to develop policy in the context of the risk assessment and risk management process we would like to have complete information on the following:

- nature and extent of the effects;
- dose/exposure response relationship(s);
- mechanism(s) of the effects;
- exposures/ambient levels;
- sources of the harmful concentrations;
- benefits of abatement;
- costs of abatement strategies.

A detailed discussion of the latter two points is outside the scope of this chapter; the abatement policies and measures discussed below have in general been established to control other pollutants as well as particles. A discussion of the costs and benefits would therefore need to be wide and will not be attempted here. Working through the remaining points for particles, it now appears to be widely accepted that there are health effects of particles at current concentrations. The early evidence for the problem came from the epidemiological work of Schwartz, Dockery and Pope in the USA, and this has been repeated in Europe and elsewhere (Department of Health, 1996). While the relative risks which these studies have found have been small (in the wider epidemiological sense), nonetheless a remarkably consistent series of associations between health effect outcomes and measures of particles have emerged. These health outcomes include measures of lung function and morbidity, such as asthma and other respiratory conditions, as well as both respiratory and all-cause mortality. This is not the place to review these studies, but reference may be made to the review of the subject by the World Health Organization (WHO) in the revision of their air quality guidelines, as yet not formally published, and the report of the UK Committee on the Medical Effects of Air Pollutants (COMEAP) on Non-Biological Particles and Health (Department of Health, 1996). These reports reviewed the evidence linking particles with human health effects. A further major assessment of the health effect evidence has been provided by the US Environmental Protection Agency (USEPA) in its Criteria Document (USEPA, 1996) which lead to the promulgation of national air quality standards for particles in the USA. The COMEAP report reviewed the published literature and concluded that it would be imprudent not to consider the association between particles and health effects as causal. There have been criticisms of the body of epidemiological studies, but by and large, where challenges have been mounted, and independent 'arbitration' studies have been carried out, the original associations have held up to scrutiny. It seems reasonable to conclude, therefore, that there is a broad consensus that there are associations between health outcomes and particle concentrations which require a policy response. It is worth noting that, to date, the great majority of these studies have reported associations between health outcomes and particles on a daily basis. Few studies have considered long-term exposures. Where this has been done, the magnitude of the effect is significantly greater than in the daily time-series studies, and further work in this area is needed.

The question then arises as to the exposure–response relationships, that is, the magnitude of health outcomes in relation to given concentrations, or increases in concentration, of particles. The WHO assessment felt unable to recommend an air quality guideline for particles. The WHO nonetheless derived exposure–response relationships from the published literature, for both PM_{10} and $PM_{2.5}$, which took into account not only the original work in the USA, but also the results of the more recent European epidemiological studies on particles. The Commission of the European Communities used this assessment as part of its preparation for a proposed Directive on a series of pollutants including particulate matter, and as part of this exercise published a Position Paper from a Working Group on Particles, which set out the risk assessment evidence for the effects on human health, incorporating the WHO exposure–response relationships.

Having reached this stage in the policy assessment, the next question which arises is to ask, what it is that one should regulate, control or abate? It is here that the specific difficulties relating to atmospheric particles begin to emerge in the context of the third, fourth and fifth bullet points in the list above. With a 'normal' pollutant, such as SO_2 or NO_2, the main problem at this stage is deciding which sources or source categories make the important contributions to harmful levels. This is often a straightforward exercise in emission inventory construction and dispersion modelling, but in the case of particles further complications arise. The measure known as PM_{10} is clearly defined in terms of the size distribution of particles it contains, but the nature of these particles and their sources can range very widely indeed.

The source areas of a sample of PM_{10} measured in an urban area in the UK for example, can range from local emissions in and around the urban area, from the rest of the UK, to the rest of Europe and beyond. This is also the case for other primary pollutants which air quality management schemes have to deal with. A further complication relating to particles is that the so-called secondary particles (sulphates, nitrates, organic particles, etc.) are formed in moving air masses and precursor emissions from 100s or even 1000s of kilometres distant can be important. In this regard, secondary particles are no different from ozone or acid deposition in an air quality management sense, in that transboundary management methods are required for their control. Where particles differ uniquely from the other pollutants is that, while a molecule of ozone formed from 'local' precursor emissions is identical to another formed from distant transboundary emissions, this is not the case for particles. Local primary combustion emissions are quite different in nature (chemical and physical) from secondary particles formed by atmospheric chemical and physical reactions. The difficulty is that while the health effect studies have established a link between effects and broad measures of particles, they have not as yet been able to define the harmful component or components of the particle distribution, in either chemical or physical (i.e. a more specific size fraction) terms. Most research in the field of particles and health is currently engaged in this problem.

Although the epidemiology has found associations between PM_{10} and health effects, from this conclusion alone it is not obvious which are the harmful components of the PM_{10} requiring control. This is complicated by the fact that associations have also been found between health outcomes and other measures of particles. There is emerging epidemiological evidence that the finer fraction of PM_{10} is the most important, and this is supported by considerations of the extent of penetration and deposition in the deeper extremes of the respiratory system. There is as yet little firm evidence in Europe that fractions other than PM_{10} are important, although the epidemiological evidence on $PM_{2.5}$ is growing.

There have also been suggestions in the literature that metrics other than PM_{10} or $PM_{2.5}$ are important, most notably total numbers of particles, rather than the mass of any one size fraction. This is discussed below in the section dealing with mechanisms, but there is some evidence (Peters *et al.*, 1997) that the *number* of so-called ultrafine particles (less than about 0.1 μm in diameter) shows better correlations with reductions in peak expiratory flow than do mass measurements of PM_{10} and of combined fine and ultrafine particles.

In attempting to determine which aspect of the particle mix to control, it is also very important to consider the relevant averaging time of the concentrations in question. This is most important in devising cost-effective control strategies since, for example, a tall stack may be solely responsible for short period peak concentrations above some threshold of harm, but might well make a negligible contribution to a long period average concentration should that be of relevance to adverse health effects. In the case of particles, virtually all the evidence to date has come from time series studies using 24 h average concentrations, with a variety of lag periods of the order of one to several days. From these studies one might conclude that the relevant averaging time for harmful effects is no longer than, and possibly shorter than several days. There are a small number of studies which have investigated longer term exposures, which show larger effects (per unit concentration) than the time series studies, but as yet there is not as large a body of evidence linking annual average exposures with effects, as there is for the daily time series studies.

From an air quality management point of view, measures of particles finer than PM_{10} would be easier to handle, since there are significant contributions to PM_{10} from the coarse fraction above 2.5 μm arising from natural sources, such as wind-blown dusts and soils, or from man-influenced sources which are difficult to control, such as re-suspended road dust and construction materials. However, even in the event that consensus is reached that the fine fraction (less than 2.5 μm) or the ultrafine fraction, should be regulated, until there is firm evidence that the toxic or harmful component(s) of PM_{10} do *not* reside in the coarse fraction, it seems imprudent to move away from the regulation of PM_{10}. Indeed, this is the case in the USA, where as well as promulgating a new air quality standard for $PM_{2.5}$, the USEPA have retained air quality standards for PM_{10}.

11.3 Mechanisms of effects of particles on health

One can view the establishment of the association between particles and effects on health through the body of epidemiological studies referred to above as a first phase, with the move towards a search for one or more mechanisms of action as a second phase. This second phase has seen a very large increase in activity in the last 2 or 3 years, and the field continues to advance rapidly. It is fair to say that, while no single mechanism or toxic component has been identified unequivocally, there is now, through the mechanistic studies which have been carried out, a much firmer basis of biological plausibility on which to interpret the epidemiological studies.

Most of the recent work has focused around ultrafine particles in terms of their physical size and their component chemical composition. The early work of Oberdörster (Oberdörster *et al.*, 1992) showed that while larger particles (0.1–2.5 μm) of substances such as TiO_2 were not harmful to rats, ultrafine particles of the same compound produced unexpectedly toxic effects. Following this early work, in 1995 two papers appeared in the literature suggesting possible mechanisms of action of ultrafine particles (Oberdörster *et al.*, 1995; Seaton *et al.*, 1995). These hypotheses involve the concept of the numbers of ultrafine particles being a key component of the process without reference to their chemical composition.

For ultrafine particles, even if the mass dose is relatively small, the numbers of particles can be very large indeed (since for a given total mass of a collection of particles of equal densities, the number will be proportional to the inverse of the cube of the radius. For equal masses of 10 μm and 0.01 μm particles, there will be 10^9 more particles of the smaller size.) This line of thinking provides a possible explanation for the low mass-dose effects of atmospheric particles, and much research is currently underway to test these and other mechanisms.

Other suggestions have involved the chemical components of particles. A series of papers by Costa and co-workers at the USEPA (Dreher *et al.*, 1997) have studied the effects of residual oil fly-ash particles (mass median diameter around 2 μm) on rats and concluded that the transition metal content is important, particularly iron, nickel and vanadium. Donaldson and colleagues (Zhang *et al.*, 1998) have also investigated the effects of ultrafines on rats in relation to some chemical components, and have shown that ultrafine nickel was significantly more toxic than ultrafine cobalt or titanium dioxide. Studies by this group and others have speculated further on how these components might act and hypotheses involving free radical generation and redox reactions involving transition metals have been suggested.

A recent review of the mechanisms of action of ultrafine particles on lung injury has been given by Donaldson (Donaldson *et al.*, 1998), but the field is moving very rapidly and new developments will certainly build on even this recent paper. As noted above however, the important point for the development of air quality management plans and policies, is that there is now a much firmer basis of biological plausibility on which to interpret the epidemiological studies than there was when those studies initially appeared, even though the active component or 'silver bullet' has not yet been identified unequivocally.

11.4 Measurement methods for particles

A crucial component of any air quality management strategy, but one which is often overlooked, is the measurement method for the pollutant in question. The reason that considerations of measurement have not figured highly for gaseous pollutants, is that there is generally little dispute over standard methods of measurement and their calibration. This is not the case for particles. The fundamental reason is that the term 'particles', or even a more precise term such as 'PM_{10}', could in principle, and generally does in practice, cover a wide and variable range of components. There is therefore not an absolute standard for a collection of particles in air, in the sense that a standard sample of, say, SO_2 could be prepared by an accredited standards laboratory to a known concentration, so that air pollution analysers could be calibrated and standardized.

There will therefore always be an element of arbitrariness in measurement standards and methods for particles, in the sense that in the absence of an absolute standard, there will have to be consensus on a definition, not only of the particle size distribution of interest, but also of the 'standard' or 'reference' method used to measure it. The choice of definitions of size distributions and measurement methods is not of course, totally arbitrary, but has generally up to now been based on physiological principles. For example, the size fraction known as 'PM_{10}' adopted originally by the USEPA, was based on the concept of the 'thoracic fraction' of airborne particles, that is, those which would penetrate beyond the larynx in the human respiratory system, and were thought to constitute the size fraction of most concern for health.

One of the problems faced in Europe has been the lack of a modern standard reference method for particles. There is an existing British Standard method for black smoke (BS 1747) but this dates from 1961 and is not a gravimetric method, nor does it collect a sample with a well-defined size-fraction. Indeed it was only in June 1998 that CEN, the European standards organization, adopted a reference method for PM_{10} sampling known as CEN 12341. This lack of standardization has posed problems, not just for policy makers, but for the science base in Europe, in that data on airborne particles (other than black smoke), where they exist at all to date, have been measured with a variety of methods, with the consequent difficulties in comparability. This is one area where the policy drivers have been ahead of the science base. The policy process, exemplified by the recently agreed EU Air Quality Daughter Directive, has coped with this difficulty by incorporating the CEN method as the reference method, but allowing Member States to use other methods, providing that acceptable relationships with the reference method can be established.

What is more, as has been discussed above, there are emerging health grounds for preferring measures of particulate matter other than PM_{10}. Again the atmospheric measurement base could be considered to be behind this policy push. In Europe there are very few data on $PM_{2.5}$ or other measures of particles finer than PM_{10}, and there are no standard reference methods for any such measures. CEN is expected to begin work on the development of a reference method for $PM_{2.5}$, and many Member States of the EU have begun to make measurements of $PM_{2.5}$, and other metrics, notably number count. It will nonetheless be some years before sufficient data have been collected, and experience gained with the new measurement methods, so that the legislative process can begin to incorporate new metrics, if that is seen as appropriate based on the developing health effect evidence. The process of negotiation of the recent EU Directive demonstrated commendable foresight and flexibility in adapting to this state of affairs by requiring Member States to undertake measurements of $PM_{2.5}$. This therefore avoids the circular argument which runs 'we can't have a limit value for X because we have no data; we have no data because there is no limit value requiring measurements'.

When the Directive is reviewed in 2003 there should therefore be not only a reference method for $PM_{2.5}$, but also a reasonably extensive database of $PM_{2.5}$ concentrations in Europe.

Reverting to PM_{10}, there are still problems arising from the multiplicity of measurement methods in use in Europe. One of the most obvious is the fact that in the past few years, evidence has begun to emerge that some continuous methods are designed in such a way as to suffer from inherent under-reading when compared with the gravimetric reference methods. The two methods which have been most extensively investigated in this regard are the so-called TEOM and the beta-ray attenuation method, both of which are used extensively in Europe. In the case of the TEOM method, the reason for the under-reading is well known. The sample inlet line is heated (to 50°C) to prevent condensation and this results in losses of the more volatile particles such as nitrates, and the more volatile organic compounds. The extent of the differences between these methods and the gravimetric methods will depend on the nature and location of the measurement site, and the Commission and the Member States using these methods will now begin the task of seeking to establish relationships between these methods and the reference methods. It is worth noting that the gravimetric reference methods may also lose volatile materials, depending on the conditions in which the filters are handled. Lest there be any misconception over the usefulness of the TEOM and beta-ray methods from the preceding discussion, it should be noted that these continuous methods enable short-period concentrations (e.g. hourly or shorter) to be measured in real time. This is invaluable in the research activities attempting to identify sources of atmospheric particles, as hourly correlations with other pollutants from co-located measurements can lead to a wealth of information. The gravimetric reference methods are 'off-line' methods and are usually coarser in time resolution in the sense that filter samples have to be collected over period usually of 24 h in operational networks, and samples weighed at a central laboratory. Moreover, if the measuring organization wishes to disseminate pollution data to the public, then the continuous automatic methods are the only solution. In both cases a trade-off is made between the value of the high-resolution, real-time data and the possibility of under-reading of the measurement method.

11.5 Source apportionment

It was noted in an earlier section that one of the fundamental requirements of an acceptable air quality management policy is the knowledge of the sources of the harmful concentration levels. Again, like other aspects of the particle problem, this causes particular difficulties compared with the gaseous pollutants. Not only is there a wide range of man-made sources of particles, there are also, particularly for PM_{10}, significant natural sources, as well as man-influenced sources which are difficult, if not impossible, to control.

Atmospheric levels of PM_{10} can be considered as being composed of three main categories. Primary particles are emitted directly by combustion processes, and are generally less than 2.5 μm and often less than 1 μm. Secondary particles are those which are formed in the atmosphere from chemical reactions and include sulphates and nitrates formed from the reactions of emissions of SO_2 and NO_x; these are also generally less than 2.5 μm, but the size could vary depending on humidity. The third category are the so-called coarse particles which are formed from a variety of non-combustion sources. These include natural events such as wind-blown dusts and soils, forest and other natural fires, and man-influenced sources such as resuspended road dust and tyre debris, and construction and mining/quarrying activity. The particles generated by these sources mostly arise from mechanical attrition and are thus relatively large, that is they are generally greater than 2.5 μm. Biological particles such as pollens and fungal spores tend to be relatively large, but can be found in the PM_{10} fraction. The extent to which such species influence effects on health is still a matter for research.

This categorization of the sources of atmospheric particles is convenient from an air quality management point of view in that the *emission* source categories of the primary, secondary and coarse particle fractions are generally different. There will clearly be exceptions in many cases, but very broadly, in a given urban area, the primary particles will arise from combustion sources in and around the urban area with a relatively small degree of long range transport from more distant areas; the secondary particles could however have a significant proportion of long-range transported material since the particle size and deposition velocities of sulphate and nitrate particles are small and their atmospheric lifetimes are long. The lifetimes of the coarse particles are short and their atmospheric transport distances are also short. Coarse particle concentrations in an urban area will therefore arise predominantly from within that area.

In designing a rational control policy for atmospheric particles, it is important to attempt to quantify as far as possible the contributions of the different source categories to atmospheric levels. This is not a straightforward task, but some significant progress has been made in the last year or so in the UK (Stedman, 1998). This receptor modelling work has involved a regression technique for apportioning the measured PM_{10} data on independent variables of black smoke or NO_x (as a surrogate for primary particles) and sulphate concentrations (as a surrogate for secondary particles). The 'constant' term in the linear regression represents the other source categories which are considered to be principally the coarse fraction of the PM_{10} referred to above. The method has been described more fully in the published paper. Since the method seeks to apportion the measured PM_{10} data, it is likely to be more accurate-certainly for annual means-than a 'bottom up' calculation from a deterministic dispersion model which would not only suffer from the inherent uncertainties of any atmospheric model, but would also have to use emission inventories whose coverage for all components of PM_{10} is incomplete and whose accuracy is difficult to quantify. An example of the results obtained by this receptor modelling method is given in *Table 11.1*. As a very broad rule of thumb, the method suggests that annual mean PM_{10} concentrations in UK urban areas are composed of roughly equal proportions of primary, secondary and coarse particles. The proportions will of course vary day by day, and even hour by hour.

This step forward allows estimates to be made of the effectiveness of different abatement strategies. With additional information from emission inventories, numerical modelling and an understanding of atmospheric processes, the changes in these classes of particles can be estimated. This is not a simple task, and the work of the APEG has described this process in more detail than can be accommodated here. As a first approximate attack on the problem, local primary emissions in future years can be modelled with urban scale dispersion models such as the standard Gaussian plume models, or more advanced models with better treatments of boundary layer turbulence. Primary emissions arising from more distant sources can be treated using models designed to deal with the appropriate spatial scales of transport and dispersion.

Secondary particles are somewhat more difficult to treat, in that their concentrations are determined by atmospheric chemistry as well as dispersion and transport. The timescales of these processes mean that the secondary particle problem is a regional and international transboundary problem, similar to that of ground level ozone. Also of importance are the dynamics of particle accumulation and agglomeration and other processes which not only determine particle size distributions and hence deposition velocities and transport distances, but also the chemical nature of the particle distribution. Some initial steps have been made in modelling secondary particles on a European scale, notably by the EMEP Meteorological Synthesizing Centre-West in Norway and the

Table 11.1. Receptor modelling source apportionment of PM_{10} in some UK cities for 1995 and 1996 ($\mu g\ m^{-3}$)

Site	1995			1996		
	Mean primary contrib.	Mean secondary contrib.	Mean 'other' contrib.	Mean primary contrib.	Mean secondary contrib.	Mean 'other' contrib.
London, Bloomsbury	9	9	11	9	10	11
Birmingham, centre	7	8	8	7	9	8
Bristol, centre	7	8	11	7	8	11
Manchester, Piccadilly	–	–	–	8	8	10
Newcastle, centre	6	8	7	6	10	8
Belfast, centre	9	6	9	9	6	9

EURAD model in Germany (Ackermann *et al.*, 1998) which incorporates detailed particle dynamics algorithms. Despite this progress, there is still much work to do before we are in a position to model secondary particles in deterministic mechanistic models with confidence. Nonetheless, there is a sufficient amount known about the nature and formation of such particles, to enable approximate estimates of future concentrations to be made. For example, the bulk of the secondary particles in North-West Europe are composed of sulphates and nitrates, mainly of ammonium. The EMEP model has been used to predict future ammonium sulphate and nitrate concentrations on the basis of current policies for the reduction of SO_2 and NO_x emissions, suggesting that between 1996 and 2010 this measure of secondary particles should reduce by about 30%.

11.6 The current policy position

In the past few years there has been activity in the UK and in Europe to develop policy on atmospheric particles. In the UK, the Expert Panel on Air Quality Standards in 1995 recommended a standard of 50 μg m^{-3} as a running 24 h average PM_{10} concentration, noting that the standard should apply to measurements of the same type as were made in the UK national monitoring network. In 1997, this recommendation for PM_{10} was incorporated into the National Air Quality Strategy (NAQS), along with standards for other pollutants. Moreover, as well as setting out standards, which represent benchmarks below which one would expect little or no effect on health from the pollutant in question, the NAQS also incorporated objectives of policy for each pollutant. These objectives incorporated considerations of technical feasibility, costs and benefits, and in many cases were expressed as a certain percentile level of achievement of the standard itself. For PM_{10}, the objective was 50 μg m^{-3} as a daily maximum running 24 h mean to be achieved on all but 1% of days in a year.

The present government, in July 1998, accepted the Strategy published by the previous administration, but decided to bring forward by a year the timing of the Review of the Strategy, to completion by the end of 1998. This was duly accomplished and a report on this review was published for consultation in January 1999 (DETR, 1999). In the course of this review, the quantification of the contribution of primary and secondary particles to measured PM_{10} levels improved considerably, as did the quantification of the role of transboundary transport. It became clear that the original objective would not be achievable on the basis of controls in the UK alone, so the report made a proposal that the original objective be replaced by the PM_{10} Stage 1 limit value in the EC Daughter Directive (see below).

In the European Union, the so-called Air Quality Framework Directive 96/62/EC was agreed in 1966, which set out the overall framework for ambient air quality management in the EU, and drew up a list of pollutants, including particulate matter, for which the European Commission would come forward with proposals for Directives. The first of these so-called Daughter Directives reached Common Position in June 1998. This contains limit values for PM_{10} as well as for SO_2, NO_2 and lead, and these are shown in *Table 11.2*. The Stage 2 limit values are more stringent than those of Stage 1, and the Directive provides for a review in 2003 of the provisions of the Directive and the experiences of Member States in implementing the Directive in the intervening period. Moreover, as has already been mentioned, the Directive also requires Member States to make measurements of $PM_{2.5}$, so that an EU-wide database of $PM_{2.5}$ concentrations should emerge as a result of this exercise. The European standards organization, CEN, has recently agreed a reference method for the measurement of PM_{10}, and is expected to begin work on such a method for $PM_{2.5}$.

There has also been considerable activity over the past few years on developing policies to reduce emissions of primary particulate matter, and the precursors of secondary particles. Of measures to reduce primary particle emissions, the most important have been the EU Directives to reduce emissions of particles from motor vehicles, and the associated Directives on fuel quality. For particles, the most important vehicles are diesels, and a Common Position on emission limits for diesel cars was agreed in 1998. More recently, in December 1998, a Common Position was reached on emission limits for heavy duty diesels for 2000 and 2005, as shown in *Table 11.3*. The limits for 2005 represent a significant tightening compared with present limits, and effectively require the use

Table 11.2. EC Directive Limit Values for Particulate Matter (PM_{10})

	Averaging period	Limit value	Date by which limit value is to be met
Stage 1			
1. 24 h limit value for the protection of human health	24 h	50 µg m^{-3} PM_{10}, not to be exceeded more than 35 times a calendar year.	1 January 2005
2. Annual limit value for the protection of human health	Calendar year	40 µg m^{-3} PM_{10}	1 January 2005
Stage 2[a]			
1. 24 h limit value for the protection of human health.	24 h	50 µg m^{-3} PM_{10}, not to be exceeded more than 7 times a calendar year.	1 January 2010
2. Annual limit value for the protection of human health.	Calendar year	20 µg m^{-3} PM_{10}	1 January 2010

[a]Indicative limit values to be reviewed in the light of further information on health and environmental effects, technical feasibility and experience in the application of Stage 1 limit values in the Member States.

Table 11.3. Emission limit levels for new heavy diesels[a]

Pollutant	CO	HC	NO_x	PM
1 April 1991 (Pre-Euro 1) 88/77/EEC	12.3[b]	2.6	15.8	–
1 October 1993 (Euro 1) 91/542/EEC	4.9	1.23	9.0	0.4
1 October 1996 (Euro 2) 91/542/EEC	4.0	1.1	7	0.15[c]
[1 October 2001 (Euro 3)][d]	2.1	0.66	5.0	0.10[c]
[1 October 2006 (Euro 4)][d]	1.5	0.46	3.5[e]	0.02

[a]Diesel engined vehicles over 3.5 tonnes.
[b]Grammes per kilowatt hour (g kWh^{-1}) – a measurement related to engine power.
[c]0.25 g kWh^{-1} until 30/9/97 and 0.13 g kWh^{-1} from 1/10/01 to 30/9/06 for certain of the smallest engines in this range.
[d]Common position on EU Commission proposal agreed by EU Environmental Ministers 21 December 1998. Subject to European Parliament Agreement.
[e]From 1/10/09 the NO_x limit to be 2.0 g kWh^{-1}.

Separate limits, as follows, were agreed by the EU Environment Council on 21 December 1998 (also subject to agreement by the European Parliament) for advanced diesel and gas engines based on a new "transient test cycle" which is designed better to reflect on-road driving conditions. These are as follows:

Date	CO	NMHC[a]	M[b/c]	NO_x	PM[d]	
1 October 2001	5.45[e]	0.78	1.6	5.0	0.16	0.21[f]
1 October 2006	4.0	0.55	1.1	3.5[g]	0.03	

[a]Grammes per kilowatt hour (g kWh^{-1}) – a measurement related to engine power.
[b]Non methane hydrocarbons.
[c]Methane.
[d]natural gas engines only.
[e]diesel engines only.
[f]for certain of the smallest engines in this range.
[g]From 1/10/09 the NO_x limit to be 2.0 g kWh^{-1}.

of particulate traps. This is of particular importance given the results of some recent studies of the emissions of particles measured as number counts, which suggest that successively stringent *mass* emission limits agreed over the past decade or so, may not have reduced the *number* of particles emitted, a finding which may be of some significance for human health. However, a small number of studies further suggest that traps reduce the emissions of both mass and number of particles by large amounts. Given the uncertainties over the role of numbers of particles as opposed to mass concentrations as the active agents in the effects of particles on human health, this recent agreement is a good example of science and engineering combining to provide a 'no regrets' policy for the control of particle emissions from the heavy duty diesel sector. The extent to which similar technologies will be applicable to other sectors will no doubt be issues for debate in the near future.

There are also policy measures in place to control secondary particles, although it is fair to say that the initial driving forces behind these have been other air pollution problems such as acidification, photochemical ozone and other gaseous urban pollutants such as nitrogen dioxide, carbon monoxide and sulphur dioxide. The majority of agreements in this area have been through Protocols under the UNECE Convention on Long Range Transboundary Air Pollution. The recent Second Sulphur Protocol agreed in Oslo in 1994 commits the UK to a reduction of 80% in sulphur emissions by 2010 on a 1980 base. The NO_x Protocol agreed in 1988, returned Parties' emissions back to 1987 levels by 1994, and this Protocol is currently being revised. Considering the central role played by NO_x the revised Protocol will take into account not only acidification and eutrophication, but also ground level ozone, and will incorporate obligations on signatories to meet emission levels for sulphur, VOCs and ammonia as well as NO_x. Although the integrated assessment modelling is using targets for acidification, eutrophication and ozone to identify optimal abatement strategies, through this measure, the main sources of secondary particles will be controlled in ECE Europe, a region covering not only the current EU, but also Eastern Europe including the Russian Federation.

Similar policy measures are also being considered in the EU. The Commission is expected to produce a proposal for a National Emission Ceilings Directive which will be based on very similar integrated assessment modelling to the UNECE Protocol, but which will cover the EU Member States. The Large Combustion Plant Directive, already under discussion in the EU, also controls emissions of sulphur dioxide and oxides of nitrogen from large plants, which can lead to the formation of secondary particles, as can the VOC emissions controlled under the Solvents Directive, also under discussion in the EU. The Directive on the Sulphur Content of Liquid Fuels, which reached common position in 1998, will also reduce sulphur dioxide emissions. Although much of the attention on secondary particle precursors is directed towards stationary sources such as power plants and other industrial combustion, mobile sources are also significant sources of secondary particle precursors, especially through the emissions of NO_x and of VOCs. The substantial reductions in emissions of gaseous pollutants from this source through the fitting of three-way-catalysts (since 1993 in the EU) and the more recent 'Auto-Oil' Directives, will also help to reduce secondary particle formation.

The above discussion is a very brief account of the current policy position in the UK and Europe on particles. Measures are in place to cover virtually all the known, controllable sources of particles. However, the medical research into the health effects is moving rapidly, as is the science behind the atmospheric behaviour of particles and our knowledge of their sources from a wide variety of activities and technologies. The above account will doubtless become dated very quickly, but the need for high quality scientific, medical, and technological input to the policy process will certainly remain for some time to come.

Disclaimer

The views expressed in this paper are those of the author alone and do not represent those of the Department of Environment, Transport and the Regions.

References

Ackermann, I.J., Hass, H., Memmesheimer, M., Ebel, A., Binkowski, F.S. and **Shankar, U.** (1998) Model aerosol dynamics model for Europe: development and first applications. *Atmos. Environ.* **32** (17): 2981–2999.

Airborne Particles Expert Group (1999) *Source Apportionment of Airborne Particulate Matter in the United Kingdom.* DETR, London (also available on http://www.environment.detr.gov.uk/airq).

Department of Health (1996) *Committee on the Medical Effects of Air Pollutants, Non-Biological Particles and Health.* HMSO, London.

Department of Trade and Industry (1997) *The Use of Scientific Advice in Policy Making.* London, UK.

DETR (1999) *Review of the United Kingdom National Air Quality Strategy.* DETR, London (also available on http://www.environment.detr.gov.uk/airq).

Donaldson K., Li, X.Y. and MacNee, W. (1998) Ultrafine (nanometre) particle mediated lung injury. *J. Aerosol. Sci.* **5/6**: 553–560.

Dreher, K.L., Jaskot, R.H., Lehmann, J.R., Richards, J.H., McGee, J.K., Ghio, A. and Costa, D.L. (1997) Soluble transition metals mediate residual oil fly ash induced lung injury. *J. Toxicol Environ. Health* **50**: 285–305.

Oberdörster, G., Ferin, J., Gelin, R. Soderholm, S.C., Finkelstein, J. (1992) Role of alveolar macrophage in lung injury: studies with ultrafine particles. *Environ. Health Perspectives* **97**: 193–199.

Oberdörster, G., Gelein, R.M., Ferin, J. and Weiss, B. (1995) Association of particulate air pollution and acute mortality; involvement of ultrafine particles? *Inhalation Toxicol.* **7**: 111–124.

Peters, A., Wichmann, H.E., Tuch, T., Heinrich, J., and Heyder, J. (1997) Respiratory effects are associated with the number of ultrafine particles. *Am. J. Crit. Care Med.* **155**: 1376–1383.

Seaton, A., MacNee, W., Donaldson, K. and Godden, D. (1995) Particulate air pollution and acute health effects. *Lancet* **345**: 176–178.

Stedman, J.R. (1998) The secondary particle contribution to elevated PM_{10} concentrations in the UK. *Clean Air* **28** (3): 87–93 (also available on http://www.environment.detr.gov.uk/airq).

Tarrason, L. and Tsyro, L. (1998) Long range transport of fine secondary particles, as presently estimated by the EMEP Lagrangian model. *EMEP/MSC-W Report 2/98,* Blindern, Norway.

US Environment Protection Agency (1996) Air quality criteria for particulate matter. *EPA/600/P-95/001aF,bF &cF* (three volumes).

Zhang, Q., Kusaka, Y., Sato, K., Nakakuki, K., Kohyama, N. and Donaldson, K. (1998) Differences in the extent of inflammation caused by intratracheal exposure on three ultrafine metals: role of free radicals. *J. Toxicol. Environ. Health Part A,* **53**: 423–438.

Glossary of terms and abbreviations related to particles

AAS	Atomic Absorption Spectroscopy
Accumulation mode	A part of the spectrum of airborne particles, between approximately 0.1 and 2 μm, diameter, in which particles have a long atmospheric lifetime
Aeroallergens	Airborne antigens that cause allergic response in sensitive individuals, e.g., pollen, house dust mite
Aerodynamic diameter	The aerodynamic diameter of a particle is the diameter which the particle would have if it were spherical in shape, of unit density and to have the same sedimentation rate as at present
Aerosol	An atmosphere containing particles which remain airborne for a reasonable length of time
AI	Alveolar–interstitial: term used in deposition models to describe peripheral site of deposition of inhaled particles
Alveolar macrophages	Phagocytic scavenger cells of the tissues as opposed to those of the blood, secrete cytokines and proteases
Ambient level	The level of pollutant in the general ground level atmosphere, i.e. to which people may be exposed
BAL	Bronchoalveolar lavage
bb	Bronchiolar } Used to describe sites of inhaled particles in deposition models
BB	Bronchial } Used to describe sites of inhaled particles in deposition models
BHR	Bronchial hyperreactivity
Black Smoke (BS)	Non-reflective (dark) particulate matter, associated with the smoke stain measurement method
Bronchodilators	An agent that increases the diameter of the air passages of the lungs
Brownian motion	Constant small movement of suspended bodies due to bombardment by surrounding molecules

Cg	Gas concentration
Coarse particle mode	A part of the size spectrum of airborne particles, greater than about 2 μm diameter, in which particles have arisen mostly from disintegration of bulk solid and liquid materials
Coarse particles	Particles within the coarse particle mode.
COPD	Chronic obstructive pulmonary disease (often taken to mean chronic bronchitis and emphysema)
Count Median Diameter (CMD)	Term used to characterise the size distribution of particles in an aerosol. Fifty percent of particles are of diameter less than the count median diameter
Cp	Particle concentration
Cyclone	Device to separate particles by size involving a rotating vortex
C5a	The biologically active fragment of the discrete protein C5 which is a component of the complex series of plasma enzymes known as the Complement System
Da	A unit of mass, the Dalton, one-twelfth the mass of the nuclide of carbon-12, equivalent to 1.657×10^{-24} g
DL_{co}	Diffusing capacity of the lung for carbon monoxide
Dyspnoea	Difficult or laboured breathing
ET	Extrathoracic: parts of the air passages (nose and mouth, pharynx, larynx) outside the thorax
FEV_1	The volume of air expired during the first second of a maximal or 'forced' expiration
Fine particles	Particles smaller than about 2 μm diameter which arise mainly from condensation of hot vapours and chemically-driven gas to particle conversion processes. Often refers to the $PM_{2.5}$ fraction q.v.
Fly ash	A fine ash produced, e.g. when pulverised fuel is burned in power stations. Particle diameter 1–20 μm
FP	Fine particles *see* $PM_{2.5}$
FVC	Forced expiratory vital capacity. The volume of air expired in a forced expiration following maximum inspiration
Geometric standard deviation (σg)	Term used to describe size distribution of particles conforming to a log-normal function
G10	10th generation of branches of the airways
H^+	Hydrogen ion
Hygroscopic growth	Growth of particles due to uptake of water from the atmosphere

ICRP	International Commission for Radiological Protection
Inhalable particles	Particles which may be breathed in. 'Inhalability' is defined technically as the orientation-averaged aspiration efficiency for the human head. (Also termed inspirable)
kPa	KiloPascal: a unit of pressure. A Pascal = 1 Newton per square metre (100 KPa = 1 atmosphere = 760 mmHg)
LDH	Lactate dehydrogenase: an enzyme frequently used as a marker of cell damage
LRT	Lower respiratory tract
LUDEP	Computer model predicting regional deposition of aerosols across populations
Mass concentration	The concentration of particles in air expressed as mass per unit volume, e.g. $\mu g/m^3$
Mast cells	Cells which when cross-linked by antigen to IgE mediate an immune response. These cells release histamine and cytokines
Methacholine	A synthetic cholinergic agent with actions similar to those of acetylcholine
Micrometre (μm)	A thousandth of a millimetre, also known as a micron
MMAD	Mass Median Aerodynamic Diameter. Term used to characterise the distribution of sizes of particles in an aerosol. MMAD is the size where half the mass of the cloud is contained in particles smaller than the stated aerodynamic diameter and half in larger particles
MMD	Mass Median Diameter. As for MMAD except that the diameter considered is the actual diameter rather than the aerodynamic diameter
MMEF	Maximum mid-expiratory flow
MMMFs	Man-made mineral fibres
Monodisperse	Term used to describe an aerosol containing particles of only one size
Neutrophils	Granular white blood cells with the properties of chemotaxis, adherence to immune complexes and phagocytosis: also produce antimicrobial substances including oxidants and proteases
nm (Nanometre)	A unit of linear measure equal to one-billionth of a metre, 10^{-9} metre or 10^{-3} μm
nmole	Nanomole, 10^{-9} mole. (The mole is the relative molecular mass of a compound expressed in grams)
NO	Nitric oxide
NO_2	Nitrogen dioxide

N_2O	Nitrous oxide: an anaesthetic gas
NO_3^-	Nitrate
NO_x	Total oxides of nitrogen. Conventionally the mixture of NO and NO_2 in the atmosphere
Non-parametric methods	Statistical methods not involving assumptions regarding the pattern of distribution of the data
Nucleation mode	A part of the size spectrum of airborne particles, below about 100 nm diameter, in which particles arise mostly from fresh emissions from combustion processes, and from gas to particle conversion
Number concentration	The concentration of particles in air expressed as number of particles per unit volume
Number median diameter (NMD)	Term used to characterise the distribution of sizes of particles in an aerosol. The NMD is that size where half the total number of particles are of diameter greater than NMD and half of diameter less than NMD (same as count median diameter)
Photochemical smog	Smog caused by the formation of particles due to a chemical reaction driven by sunlight
PMNs	Polymorphonuclear neutrophil leukocytes (see neutrophils)
PM_{10}	Particulate matter less than 10 μm aerodynamic diameter (or, more strictly, the mass of particles per cubic metre which pass through a size selective inlet with a 50% efficiency cut-off at 10 μm aerodynamic diameter)
PM_{15}	As with PM_{10} but with 15 μm as the cut-off point
$PM_{2.5}$	As with PM_{10} but with 2.5 μm as the cut-off point
ppb	Parts per billion, 1 part by volume in 10^9
ppm	Parts per million, 1 part by volume in 10^6
Relative humidity (RH)	Actual vapour pressure/saturated vapour pressure expressed as a percentage. A measure of the degree of saturation of the air with water vapour
Respirable particles	Particles which can penetrate to the unciliated regions of the lung
Secondary particle or aerosol	Particles may be formed when two volatile and non-condensible vapour species react to give rise to a product with a very low vapour pressure. Such a product is described as a secondary particle to distinguish it from those arising from the reaction of liquids or solids
Smog	A term used to describe a mixture of smoke and fog. Also used to describe photochemical air pollution

Smoke	Particulate matter, <15 μm diameter, derived from the incomplete combustion of fuels
SO_2	Sulphur dioxide
SO_4^{2-}	Sulphate ion
SO_x	Oxides of sulphur: used, as is NO_x, to denote the mixture of oxides of sulphur that occur in the air
SPM (Suspended particulate matter)	A general term embracing all airborne particles
Summertime haze	Atmospheric haze produced during photochemical episodes: comprises secondary aerosols
TEOM	Tapered Element Oscillating Microbalance, an instrument that allows for the continuous measurement of suspended particulate matter in air
Thoracic particle mass	Describes that fraction of the particles which penetrates beyond the nasopharynx and larynx
Total suspended particulate (TSP)	A term describing the gravimetrically determined mass loading of airborne particles, most commonly associated with use of the US high volume air sampler in which particles are collected on a filter for weighing
Ultrafine particles	Particles of less than 100 nm diameter

Index